PUBLISHERS' NOTE

The series of monographs in which this title appears was introduced by the publishers in 1957, under the General Editorship of Dr Maurice G. Kendall. Since that date, more than twenty volumes have been issued, and in 1966 the Editorship passed to Alan Stuart, D.Sc. (Econ.), Professor of Statistics, University of London.

The Series fills the need for a form of publication at moderate cost which will make accessible to a group of readers specialized studies in statistics or courses on particular statistical topics. Often, a monograph on some newly developed field would be very useful, but the subject has not reached the stage where a comprehensive treatment is possible. Considerable attention has been given to the problem of producing these books speedily and economically.

It is intended that in future the Series will include works on applications of statistics in special fields of interest, as well as theoretical studies. The publishers will be interested in approaches from any authors who have work of importance suitable for the Series.

CHARLES GRIFFIN & CO. LTD

T0204349

OTHER BOOKS ON STATISTICS AND MATHEMATICS

Descriptive brochure from Charles Griffin & Co. Ltd.

THE MATHEMATICS OF EXPERIMENTAL DESIGN

INCOMPLETE BLOCK DESIGNS AND LATIN SQUARES

S. VAJDA

Dr. Phil. (Vienna), F.S.S.
Professor of Operational Research
University of Birmingham

BEING NUMBER TWENTY-THREE OF
GRIFFIN'S STATISTICAL
MONOGRAPHS & COURSES

EDITED BY
ALAN STUART, D.Sc.(Econ.)

CHARLES GRIFFIN & COMPANY LIMITED
LONDON

Copyright © 1967
CHARLES GRIFFIN & COMPANY LIMITED
42 DRURY LANE, LONDON, W.C.2
SBN: 85264 036 6

First published . . . 1967

SET BY E. W. C. WILKINS & ASSOCIATES LTD, LONDON
PRINTED IN GREAT BRITAIN BY LATIMER TREND & CO LTD, WHITSTABLE

PREFACE

The topics with which the two companion volumes

Patterns and Configurations in Finite Spaces

and

*The Mathematics of Experimental Design : incomplete
block designs and Latin squares*

are concerned have a long history, but the main stimulus to new developments has come from the interest of statisticians in the efficient design of experiments. The origin of some of the patterns can be traced to recreational mathematics, or to problems in pure mathematics, particularly in number theory.

The statistical aspect of most of the subjects mentioned has been dealt with in many excellent textbooks, together with their analysis which leads to inferences about the effectiveness of treatments or other choices, the outcome of which is subject to stochastic variation. No such analysis is described in the present books. They contain rather the combinatorial aspects of the construction of designs, without regard to their practical application or indeed other usefulness.

In *Patterns and Configurations in Finite Spaces*, the first chapter contains the algebra which is needed in the subsequent pages; it is, of course, not a course in algebra, but only a review of certain parts of it.

Chapters II and III contain the fundamentals of finite geometry, two-dimensional and more-dimensional geometry being treated in separate chapters. In Chapter IV we reach configurations.

In *The Mathematics of Experimental Design* we start with a short review of algebraic facts, and then deal fully with some special cases of configurations treated in the first-mentioned book. Thus, balanced incomplete block designs are described in Chapter II, and partially balanced incomplete block designs in Chapters IV and V, the last chapter being concerned with a special case of designs introduced in the former, namely those of two associate classes.

Chapter III is devoted to orthogonal arrays and to a special case of these, namely Latin squares and their orthogonal sets. It contains an account óf the most recent results in this field, including the proof of the falsity of a hypothesis by Euler, which was long considered to be plausible.

The bibliographies mention only those papers which are referred to in the text of the respective book. A complete bibliography of the subject would be too extensive.

The exercises are meant to test the reader's understanding, and to serve the teacher of these topics. Both the reader and the teacher will welcome the inclusion of solutions to the exercises. Examples are also found throughout both books.

Birmingham, 1967   S. VAJDA

CONTENTS

CHAPTER I

INTRODUCTION

In this book we shall be concerned with the mathematical aspect of designs which have been developed for the purpose of analysing first agricultural, and then also other statistical experiments. The analysis of the results of such experiments is outside the scope of the book. We shall concentrate on the description and construction of the designs.

Much of the mathematics which forms the basis of the construction of designs possessing those features which make them useful for experimental layouts has been dealt with in the companion volume [129]. We start here by stating briefly, without detailed proofs, those facts which we need in the following chapters, and refer the reader for a rigorous treatment to the presentation in that volume.

Finite groups

A finite group is a set of a finite number of distinct elements, say $a_1, \ldots, a_s$, with a rule of operation for combining any two of them in a given order. The operation satisfies the following rules:

(1) If both a_i and a_j are elements of the group (not necessarily distinct) then their combination, written $a_i a_j$, and also called their *product*, is also an element of the group.

(2) Associative law: if a_i, a_j, and a_k are elements of the group (not necessarily distinct), then $(a_i a_j) a_k = a_i (a_j a_k)$.

(3) The group contains a unique "identity" element, say I, such that for every element a_i of the group we have $a_i I = I a_i = a_i$.

(4) For every element a_i of the group there exists a unique element, a_j say, such that $a_i a_j = a_j a_i = I$. We write $a_j = a_i^{-1}$. Of course, $I^{-1} = I$.

We do not necessarily assume that $a_i a_j = a_j a_i$ for any pair of elements of the group. A group in which this equality holds is called *commutative*, or *Abelian*.

The number of elements in a group is called its *order*, and the smallest integer m such that $a_i^m = I$ is called the order of a_i. Every element has an order, and it is always a factor of the order of the group.

A group which consists of all the powers of one element is called *cyclic*. It is Abelian.

Two groups are called *isomorphic* if each element of one can be made to correspond to an element of the other in such a way that the product of two elements of one group corresponds to the product of the corresponding elements of the other group. The relationship between two isomorphic groups is called an *isomorphism*, and if the two groups are, in fact, the same, then the isomorphism is called an *automorphism*.

If the product of two elements corresponding in an automorphism are multiplied, they produce again an element of the group. If the products of two such elements are all different, then the isomorphism is a "complete mapping" (see [64]). A group of order $4t + 2$ cannot have a complete mapping.

The permutations of m symbols form a group of order $m! = 1.2...m$, where m is called the *degree* of any permutation group of m symbols. A permutation group is conveniently described by a sucession of cycles of non-overlapping sets of elements. Each cycle, say $(a\,b\,c\,...\,d)$, indicates that the permutation changes a into b, b into c, ... and finally d back into a.

Examples

The permutation which changes the order 12345 into 21543 may be written as (12)(35)(4), where the last "cycle" (4) indicates that 4 remains in its original position. This cycle could remain unrecorded, thus: (12)(35).

The permutations I, (12)(34), (13)(24), (14)(23) form an Abelian (but not cyclic) group of order 4 and degree 4.

The permutations I, (1234), (13)(24), (1432) form a cyclic, and hence Abelian group of order 4 and degree 4.

The permutations

I,	(12)(34),	(13)(24),	(14)(23)
(234),	(132),	(143),	(124)
(243),	(142),	(123),	(134)

form a group of order 12 and degree 4, which is not Abelian. For instance,

$$(12)(34).(234) = (132), \text{ but } (234).(12)(34) = (124).$$

Galois fields

A Galois field is a set of s (> 1) distinct elements which form a commutative group with regard to a combination which we call, and write as, *addition*, the identity elements being denoted by 0, and the inverse of a_i by $-a_i$. The elements are also subject to another combination, called *multiplication*. With regard to the latter, $a_i 0 = 0 a_i = 0$ for all a_i. Apart from 0, the elements form a commutative group under multiplication, and the identity element of this group is denoted by 1, and the inverse of a_i by a_i^{-1}.

The elements of a Galois field satisfy also the "distributive law" $a_i(a_j + a_k) = (a_j + a_k)a_i = a_i a_j + a_i a_k$.

A Galois field exists for every s which is a power of a prime, and for no other s. All Galois fields of the same order are isomorphic as regards addition as well as multiplication, and we may therefore say that, in an abstract sense, there exists only one Galois field for every prime power. We denote it by $GF(s)$.

Example

$GF(4)$. We denote the elements by $0,1,y,z$. They satisfy the following tables of

	Addition					Multiplication			
	0	1	y	z		0	1	y	z
0	0	1	y	z	0	0	0	0	0
1	1	0	z	y	1	0	1	y	z
y	y	z	0	1	y	0	y	z	1
z	z	y	1	0	z	0	z	1	y

Finite projective geometries

A finite projective geometry of m dimensions consists of the ordered sets $(x_0, x_1, ..., x_m)$, called *points*, where the x_i are elements of the $GF(s)$ and not all $x_i = 0$. The space is denoted by $PG(m,s)$ (for "projective geometry"). A $PG(2,s)$ is also referred to as a *finite projective plane*. (Other types of such planes also exist (cf. [129].)

The x_i are homogeneous co-ordinates of the point, i.e. for any $k \neq 0$, the point $(kx_0, ..., kx_m)$ is considered to be the same as the point $(x_0, ..., x_m)$.

Any of the x_i can have s different values, but they must not all of them be simultaneously zero. Moreover, multiplying all of them by any of the s elements of the $GF(s)$, except by 0, produces the same point. It follows that the number of points in a $PG(m,s)$ is

$$(s^{m+1} - 1) / (s - 1) = V(m,0; s), \quad \text{say.}$$

The $PG(m,s)$ contains k-flats, $(k = 0,1, ..., m - 1)$, defined as the set of all points whose co-ordinates satisfy $m - k$ independent linear homogeneous equations with coefficients from the $GF(s)$, not all simultaneously 0 within the same equation.

Alternatively, a k-flat in a $PG(m,s)$ consists of all points with co-ordinates $(a_0 x_{00} + \dots + a_k x_{k0}, \dots, a_0 x_{0m} + \dots + a_k x_{km})$ with all elements from the $GF(s)$. The a_i are not simultaneously all 0, and the matrix

$$\begin{pmatrix} x_{00} & \cdots & x_{0m} \\ & \cdots & \\ x_{k0} & \cdots & x_{km} \end{pmatrix}$$

has rank $k + 1$. Each k-flat contains $V(k,0;s)$ points. In particular, there are $s + 1$ points on a 1-flat, also called a *line*.

A 0-flat is a point, and an $(m - 1)$-flat in a $PG(m,s)$ is called a *hyperplane*.

It can be shown (see [129]) that the number of k-flats in a $PG(m,s)$ equals

$$\frac{(s^{m+1} - 1)(s^m - 1) \dots (s^{m-k+1} - 1)}{(s^{k+1} - 1)(s^k - 1) \dots (s - 1)} = V(m,k;s), \text{ say.}$$

The number of t-flats which contain a given k-flat $(k \neq t)$ is

$$V(m - k - 1, m - t - 1; s) = V(m - k - 1, t - k - 1; s).$$

For instance, there are $(s^m - 1)/(s - 1)$ hyperplanes through a point, and $(s^{m-1} - 1)/(s - 1)$ through a line.

Example

The $PG(2,4)$ is a finite plane with 21 points and an equal number of lines. On each line there are 5 points, and through every point there are 5 lines. Through any two points there is precisely one line, and any two lines meet in precisely one point.

We can also define a conic in a finite projective plane, as the set of all those points whose co-ordinates satisfy a homogeneous equation of the second order. If the quadratic form is not the product of two linear forms, then the conic is called *non-degenerate*, and consists of $s + 1$ points. As regards their enumeration, see [129].

Example

A conic in the $PG(2,4)$ is, for instance, that consisting of the five points $(1, 1, z)$ $(0, 1, y)$ $(1, 1, y)$ $(1, z, y)$ $(1, z, z)$. Its equation is

$$x_0^2 + x_1^2 + y(x_0 x_1 + x_0 x_2 + x_2^2) = 0.$$

In a $PG(2, 2^t)$ the tangents of a non-degenerate conic meet in the same point, the "nucleus" of the conic. In the example above, it is the point $(0, 1, 1)$.

Finite Euclidean spaces

A Euclidean space of m dimensions, denoted by $EG(m,s)$, is the set of all points of a $PG(m,s)$ whose first co-ordinate x_0 is not 0. Because the co-ordinates of the points in a $PG(m,s)$ are homogeneous, we can take that first co-ordinate to be 1, so that the points of a $EG(m,s)$ are given by the ordered sets $(x_1, \dots, x_m)$, where the x_i are elements of the $GF(s)$.

There are s^m such points, and the number of k-flats can be shown (see [129]) to be

$$V(m, k; s) \ - \ V(m-1, k; s) \ = \ V(m-1, k-1; s) s^{m-k}.$$

For instance, there are $s(s^m - 1)/(s - 1)$ hyperplanes.

A k-flat in a $EG(m,s)$ contains s^k points. In particular, there are now s points on a line (a 1-flat).

The number of t-flats which contain a given k-flat is the same as that in a $PG(m,s)$, namely $V(m-k-1, t-k-1; s)$.

In a EG$(2,s)$ there are conics with $s+1$, with s, and with $s-1$ points which we call, respectively, *ellipses*, *parabolas*, and *hyperbolas*. They number, respectively, $s^3(s-1)^2/2$, $s^2(s^2-1)$, and $(s^5-s^3)/2$.

Difference sets, and systems of difference sets

A set of integers $d_1, \dots, d_k$ is said to be a "difference set mod v," if the $k(k-1)$ differences $d_i - d_j$ equal any non-zero value mod v the same number (say λ) of times. It follows at once that $k(k-1) = \lambda(v-1)$.

Example

$v = 7$. $1, 2, 4$ form a difference set, with $\lambda = 1$.

$v = 11$. $1, 3, 4, 5, 9$ form a difference set, with $\lambda = 2$.

We speak of a system of difference sets

$$d_{11}, \dots, d_{1k}$$
$$- \quad - \quad -$$
$$d_{m1}, \dots, d_{mk}$$

if the d_{ij} are elements of a group of order s, and such that if we compute all the $mk(k-1)$ values $d_{ij_1}^{-1} d_{ij_2}$ $(i = 1, \ldots, m; j_1 \neq j_2)$ of any two elements in the same row, tnen each of the $s-1$ non-unit elements appears the same number of times. If this number is λ, then we have $mk(k-1) = \lambda(s-1)$.

Multiplication, and reciprocity, are to be interpreted in the sense of the rule of combination of the group. If the group is that of addition, and $m = 1$, then we are back at the definition of a difference set.

Example

$k = 3$, $m = 2$, $s = 7$.   1 2 4
3 5 6.

CHAPTER II

BALANCED INCOMPLETE BLOCK DESIGNS

In this chapter we deal with a design consisting of two types of objects with an incidence relation between them. We shall call the two types of objects *varieties* and *blocks*; the latter contain the varieties. These expressions are taken from agricultural field trials, where these designs were first applied to statistical experiments. The number of varieties will, as a rule, be denoted by v, and that of the blocks by b.

No variety will appear more than once in a block, and no two blocks will be identical. Also, to exclude trivial cases, we assume $v > 1$ and $b > 1$. Every block will contain the same number (say k) of varieties, and all varieties will appear in the same number (say r) of blocks. It follows that $bk = vr$.

We assume, moreover, that any pair of varieties appears together in the same number (say λ) of blocks.

Any variety, x say, appears r times, and therefore together with $r(k-1)$ varieties in the same block, though they are, of course, not all different. Since every one of the $v-1$ other varieties appears λ times together with x, we have $r(k-1) = \lambda(v-1)$.

Such a design is called a *balanced incomplete block design* (b.i.b.d.). It is a special case of complete, or balanced, configurations (see [129]). In [131], Yates called b.i.b.d's "incomplete randomized blocks".

If $b = v$, and hence $r = k$, then the b.i.b.d. is called *symmetric*. If any two of the blocks have the same number (say μ) of varieties in common, then the b.i.b.d. is called of *linked block* type. We shall see that symmetry implies linkage, and vice versa.

A b.i.b.d. is called *resolvable* if its blocks can be grouped into sets of equal sizes, so that the blocks of each set contain between them all the varieties once; if, moreover, any two blocks from different sets have the same number of varieties in common, then the b.i.b.d. is called *affine resolvable*.

We give first a few examples.

Finite spaces

The points of a $PG(m,s)$, or of a $EG(m,s)$, considered as varieties, and its t-flats as blocks, form b.i.b.d's. The parameters of such a design are as follows:

$PG(m,s)$		$EG(m,s)$
$V(m,0;s)$	v	s^m
$V(m,t;s)$	b	$V(m,t;s) - V(m-1,t;s) =$ $V(m-1,t-1;s)s^{m-t}$
$V(m-1,t-1;s)$	r	$V(m-1,t-1;s)$
$V(t,0;s)$	k	s^t
$V(m-2,t-2;s)$ if $t \neq 1$, or 1, if $t=1$	λ	$V(m-2,t-2;s)$ if $t \neq 1$ or 1, if $t=1$.

As special cases, we mention $t = m-1$:

$PG(m,s)$		$EG(m,s)$
$(s^{m+1}-1)/(s-1)$	v	s^m
$(s^{m+1}-1)/(s-1)$	b	$s(s^m-1)/(s-1)$
$(s^m-1)/(s-1)$	r	$(s^m-1)/(s-1)$
$(s^m-1)/(s-1)$	k	s^{m-1}
$(s^{m-1}-1)/(s-1)$	λ	$(s^{m-1}-1)/(s-1)$

These designs, derived from $PG(m,s)$, are symmetric.

For special values of s we have, when $t = 1$:

$s = 2$				$s = 3$		
$PG(m,2)$		$EG(m,2)$		$PG(m,3)$		$EG(m,3)$
$2^{m+1}-1$	v	2^m		$\frac{1}{2}(3^{m+1}-1)$	v	3^m
$(2^{m+1}-1)(2^m-1)/3$	b	$2^{m-1}(2^m-1)$		$(3^{m+1}-1)(3^m-1)/16$	b	$3^m(3^m-1)/6$
2^m-1	r	2^m-1		$\frac{1}{2}(3^m-1)$	r	$\frac{1}{2}(3^m-1)$
3	k	2		4	k	3
1	λ	1		1	λ	1

For general s, $m = 2$, $t = 1$ we have:

PG(2,s)		EG(2,s)
$1 + s + s^2$	v	s^2
$1 + s + s^2$	b	$s(s + 1)$
$1 + s$	r	$1 + s$
$1 + s$	k	s
1	λ	1

Conics

From the enumeration of non-degenerate conics in a $PG(2,s)$, given in [129], it follows that if we take the points as varieties, and the non-degenerate conics as blocks, then we obtain a b.i.b.d. with the following parameters:

$$v = s^2 + s + 1, \quad b = s^5 - s^2, \quad r = s^2(s^2 - 1),$$

$$k = s + 1, \quad \lambda = s\ (s - 1).$$

If we take parabolas as blocks, and the points of a $EG(2,s)$ as varieties, then we have a b.i.b.d. with

$$v = s^2, \quad b = s^2(s^2 - 1), \quad r = s(s^2 - 1), \quad k = s, \quad \lambda = s(s - 1),$$

and if we use hyperbolas, we have

$$v = s^2, \quad b = \tfrac{1}{2} s^3(s^2 - 1), \quad r = \tfrac{1}{2}(s + 1)s(s - 1)^2,$$

$$k = s - 1, \quad \lambda = \tfrac{1}{2} s(s - 1)(s - 2).$$

More generally, using the results for non-degenerate quadrics given in [129], we obtain b.i.b.d's with

$$v = (s^{2t+1} - 1)/(s - 1), \quad b = s^{t(t+1)} \prod_{i=1}^{t} (s^{2i+1} - 1),$$

$$k = (s^{2t} - 1)/(s - 1), \quad r = s^{t(t+1)} (s^{2t} - 1) \prod_{i=1}^{t-1} (s^{2i+1} - 1)$$

$$\lambda = s^{t(t+1)} (s^{2t-1} - 1) \prod_{i=1}^{t-1} (s^{2i+1} - 1)$$

from points as varieties and quadrics as blocks in a $PG(2t,s)$ where $t \geqslant 2$, or

$$v = (s^{2t} - 1)/(s - 1), \quad b = \tfrac{1}{2} s^{t^2} (s^t + 1) \prod_{i=1}^{t-1} (s^{2i+1} - 1),$$

$$k = (s^t - 1)(s^{t-1} + 1)/(s-1), \quad r = \tfrac{1}{2} s^{t^2}(s^{t-1} + 1) \prod_{i=1}^{t-1} (s^{2i+1} - 1),$$

$$\lambda = \tfrac{1}{2} s t^2 (s^{t-1} + 1)(s^{2t-2} + s^{t-1} - s^{t-2} - 1) \prod_{i=1}^{t-2} (s^{2i+1} - 1)$$

from ruled quadrics as blocks in a $PG(2t-1, s)$, $t \geqslant 2$, and

$$v = (s^{2t} - 1)/(s-1), \quad b = \tfrac{1}{2} s^{t^2}(s^t - 1) \prod_{i=1}^{t-1} (s^{2i+1} - 1),$$

$$k = (s^t + 1)(s^{t-1} - 1)/(s-1), \quad r = \tfrac{1}{2} s^{t^2} (s^{t-1} - 1) \prod_{i=1}^{t-1}(s^{2i+1} - 1).$$

If, in the last two cases, $t = 2$, then the term $\prod_{i=1}^{t-2} s^{(2i+1} - 1)$ is replaced by 1.

If we use hyperbolas, we have $v = s^2$, $b = s^3(s-1)/2$, $r = (s+1)s(s-1)^2/2$ and $k = s - 1$.

$$\lambda = \tfrac{1}{2} s t^2 (s^{t-1} - 1)(s^{2t-2} - s^{t-1} + s^{t-2} - 1) \prod_{i=1}^{t-2}(s^{2i+1} - 1)$$

from unruled quadrics as blocks in a $PG(2t-1, s)$, $t \geqslant 2$.

Consider now a non-degenerate conic in a $PG(2, 2^t) = PG(2, s)$. Of the $s^2 + s + 1$ lines of the plane, $\binom{s+1}{2}$ meet the conic in two points and do not pass through the nucleus, $s + 1$ are tangents, and pass through the nucleus, $\binom{s}{2}$ do not have any point in common with the conic. We call these $\binom{s}{2}$ lines "v-lines".

Of the $s^2 + s + 1$ points of the plane, one is the nucleus, $s + 1$ are points of the conic, $s^2 - 1$ points are neither on the conic, nor is one of them the nucleus. We call these points "b-points."

We take the b-points as blocks (not as varieties), and the v-lines as varieties. We have then a b.i.b.d. with

$$v = \binom{s}{2}, \quad b = s^2 - 1, \quad r = s + 1, \quad k = \tfrac{1}{2} s.$$

Because any two v-lines meet in a single point, $\lambda = 1$.

This b.i.b.d. is resolvable (though, of course, not affine). This can be seen as follows.

Take a line which is not a v-line. There are on it $s - 1$ different b-points. Each of these corresponds to a block with $\tfrac{1}{2} s$ varieties, and two varieties in different blocks cannot be the same, because the line connecting two of the b-points is not a v-line. Hence all these blocks contain, between them, all varieties.

If we select any point on the conic, or the nucleus, and consider the $s + 1$ lines through it, then there pass through such a point $s + 1$ lines which are not v-lines, and thus we obtain $s + 1$ different sets of $s - 1$ blocks each, and each of these sets contains all varieties. We see also that the grouping into sets can be done in $s + 2$ different ways, because there are $s + 1$ points of the conic from which we can start, or we can start with the nucleus.

In [9] R.C.Bose has shown that resolvable b.i.b.d's exist for all parameters of the form $v = s^3 + 1$, $b = s^2 (s^2 - s + 1)$, $r = s^2$, $k = s + 1$, $\lambda = 1$, when s is a prime power. He takes as varieties the points of the curve $x_0^{s+1} + x_1^{s+1} + x_2^{s+1} = 0$ in a $PG(2,s^2)$ and as blocks the lines which are not tangents of this curve. For $s = 2$, we obtain the b.i.b.d. given on page 27.

We shall now describe two methods of obtaining b.i.b.d's from a given symmetric one, with parameters v, r, and λ.

Block section

The new design, called "residual", is obtained by omitting one block and all its varieties from all blocks. The new parameters are:

$$v' = v - k, \quad b' = v - 1, \quad r' = r, \quad k' = r - \lambda, \quad \lambda' = \lambda.$$

Block intersection

The other new design, called *derived*, is obtained by omitting one block and retaining in the other blocks only those varieties which were contained in the omitted block. The new parameters are:

$$v'' = k, \quad b'' = v - 1, \quad r'' = k - 1, \quad k'' = \lambda, \quad \lambda'' = \lambda - 1.$$

A method of construction, introduced in [4] (see also [5] and [2]), will be described by an example, taken from [22].

Varieties are denoted by a pair of numbers mod 7. n is an element which does not change when any number is added to it. Consider the blocks:

$$(1,1 \ \ 2,1 \ \ 4,1), \quad (3,1 \ \ 1,2 \ \ 5,2), \quad (6,1 \ \ 2,2 \ \ 5,2)$$
$$(5,1 \ \ 4,2 \ \ 6,2), \quad (0,1 \ \ 0,2 \ \ n,0).$$

Adding, respectively, 1, or 2, ..., or 6 to the first element of a pair and reducing mod 7, we obtain a resolvable b.i.b.d. with $v = 15$, $b = 35$, $k = 3$, $r = 7$, $\lambda = 1$.

Incidence matrix

We now introduce a way of representing a b.i.b.d., its so-called incidence matrix. This is a matrix of v rows and b columns. Its rows correspond to the varieties, and its columns to the blocks. We number the varieties and the blocks, and if the ith variety appears in the jth block, then we write into the ith row and the jth column a 1, otherwise we write there 0.

It is easily seen that the sum of entries in any row is r, and that in any column is k. The inner product of two different rows is λ, and if the b.i.b.d. is of linked block type, then the inner product of any two different columns is μ.

We denote the incidence matrix by A. Its transpose, A', is the incidence matrix of the so-called "dual" design, and if we write 1 for 0, and 0 for 1, then we obtain from A the incidence matrix of the "complementary" design.

We quote from [129] that all rows and columns of the matrix product AA' have the same total, namely rk, and that the determinant $|AA'|$ equals $rk(r-\lambda)^{v-1}$.

Because $r > \lambda$, this is positive, i.e. AA' is of rank v. The v by b matrix A cannot have smaller rank, hence

$$b \geqslant v.$$

This proof is taken from [6]. R.A. Fisher has proved this inequality in [43] by a different method.

As an example of the use of the concept of the incidence matrix we now prove that a b.i.b.d. is of linked block type if and only if it is symmetric.

Denote the square matrix of order v whose elements are all 1 by J_v and the identity matrix of order v by I_v. If $v = b$, then we have

$$AA' = (k-\lambda)I_v + \lambda J_v, \quad \text{and} \quad AJ_v = J_v A = kJ_v = rJ_v.$$

Hence $A'A = A^{-1}(AA')A = A^{-1}(k-\lambda)I_v A + \lambda A^{-1}J_v A$

$$= (k-\lambda)I_v + \lambda J_v = AA'.$$

This proves that a symmetric b.i.b.d. is of linked block type, and

$$\lambda = \mu = k(k-1)/(v-1).$$

Conversely, let a b.i.b.d. be of linked block type. Then

$$AA' = \begin{pmatrix} r\lambda..\lambda \\ \lambda r..\lambda \\ ..... \\ \lambda\lambda..r \end{pmatrix} \quad \text{and} \quad A'A = \begin{pmatrix} k\mu..\mu \\ \mu k..\mu \\ ..... \\ \mu\mu..k \end{pmatrix}$$

$$|AA'| = rk(r-\lambda)^{v-1} \quad \text{and} \quad |A'A| = [k+\mu(b-1)](k-\mu)^{b-1}.$$

Both these determinants are positive, and hence the rank of AA' is v, and that of $A'A$ is b. The rank of a product of two determinants cannot exceed that of either factor, and therefore we have $b \geqslant v$, and $v \geqslant b$, i.e.

$$v = b, \quad r = k, \quad \lambda = \mu.$$

In particular, let $\lambda = \mu = 1$. Then we have $v = b = r^2 - r + 1$, and denoting $r - 1$ by s, we have $v = b = s^2 + s + 1$, $r = k = s + 1$.

Thus the blocks and varieties of a symmetric b.i.b.d. with $\lambda = 1$ satisfy the conditions for lines and points of a finite projective plane. We have already seen that the converse is also true: the two concepts, namely that of a symmetric b.i.b.d. with $\lambda = 1$, and of a finite projective plane, are equivalent.

If the b.i.b.d. is resolvable, then we can derive $b \geqslant v + r - 1$, which is a stronger inequality than $b \geqslant v$. Take those columns of A which correspond to the blocks of one complete replication. The entries of each row add up to 1, and therefore not more than $b - r + 1$ columns can be independent. Hence $b - r + 1 \geqslant v$, and the inequality follows. It was proved by a different method in [129].

If the b.i.b.d. is affine resolvable, then we have r sets, of $n = b/r$ blocks each. Call the number of common varieties in any two blocks from different sets μ. Take one block, say B_0 and consider all its varieties, and all their replications in the other sets. In the sets which do not contain B_0, these varieties will appear $k(r-1)$ times, because every one is repeated $r - 1$ times in them. Also, since the $n(r-1)$ blocks in these sets each contain μ varieties in common with those of B_0, we have $k(r-1) = n(r-1)\mu$, i.e. $\mu = k/n = kr/b$.

Then, consider the $\binom{k}{2}$ pairs of varieties in B_0. They appear $\lambda - 1$ times in other blocks. In every one of these, $n(r-1)$ in number, there will be $\binom{\mu}{2}$ such pairs. Hence $\binom{k}{2}(\lambda - 1) = \binom{\mu}{2}n(r-1)$.

Replace n by b/r and μ by kr/b. Then we obtain, after some straightforward arithmetic, $b = v + r - 1$.

From an orthogonal matrix of order $4t$ whose elements are 1 or -1

we can construct a b.i.b.d. with parameters $v = b = 4t-1$, $r = k = 2t-1$, $\lambda = t - 1$ by a method indicated in [128].

We can change all signs in a row or in a column of an orthogonal matrix in such a way that all elements in the first row, and also in the first column, are 1. Orthogonality is not destroyed by this change. We then suppress the first row and the first column. As a consequence of the orthogonality of the original matrix each row will then contain $2t-1$ elements 1, and any two rows will have $t - 1$ pairs of 1's in the same column. If we change -1 into 0, then we have obtained an incidence matrix of a b.i.b.d. with the parameters as stated. (The procedure can be reversed, and we can construct a Hadamard matrix from a b.i.b.d. with those parameters.)

A design with such an incidence matrix is called a *Hadamard configuration*. An example is the $PG(m,2)$, with hyperplanes as blocks, and points as varieties. To see this, write t for 2^{m-1}. The complementary design, i.e. that whose incidence matrix is obtained by exchanging 0 and 1, is also called a Hadamard configuration. Its parameters are

$$v = b = 4t - 1, \quad k = r = 2t, \quad \lambda = t.$$

Existence theorems

The parameters of a b.i.b.d. satisfy the conditions $bk = vr$, $r(k-1) = \lambda(v - 1)$, and hence $b\binom{k}{2} = \lambda\binom{v}{2}$. It is therefore necessary that $r = \lambda(v - 1)/(k - 1)$ and $b = \lambda v(v - 1)/k(k - 1)$ be integers. There are cases where it can be shown that these two conditions are also sufficient for a b.i.b.d. with such parameters to exist. We shall discuss a few such cases.

When $\lambda = 1$ and $k = 2$, then we obtain $r = v - 1$ and $b = \binom{v}{2}$. In this case the blocks are all possible pairs of varieties.

When $\lambda = 1$ and $k = 3$, then $3b = vr$ and $2r = v-1$. These are the conditions for a Steiner triple system, and it is shown in [129] that they exist if and only if $v = 6t+1$ or $6t+3$, where t is a non-negative integer (cf. [96] and [68]).

When $\lambda = 2$ and $k = 3$ (a case studied by R.C. Bose in [4]), then $3b = vr$ and $2r = 2(v-1)$; hence $b = r(r+1)/3$ and $v = r+1$. Therefore either $r = 3t$, $v = 3t+1$, $b = t(3t+1)$, or $r = 3t+2$, $v = 3(t+1)$, $b = (t+1)(3t+2)$. If $t = 2u$, then $r = 6u$ or $6u+2$. These designs can be constructed by combining two systems of Steiner triples with the same number of elements, either permuting their notation, or taking non-isomorphic systems. If $t = 2u+1$, then $r = 6u+3$ or $6u+5$. These designs can also be constructed.

In these cases we can express our results by saying that it is sufficient for the existence of a b.i.b.d. that the expressions for r and b at the beginning of this section should be integers. These same conditions are also sufficient for $k = 3$ or 4, with any value of λ, and also for $k = 5$ when $\lambda = 1$, 4, or 20, with the possible exception of $v = 141$ in the first of these cases. This was proved in [49]. That for $k = 6$ and $\lambda = 1$ the integrality of the above expressions is not sufficient was shown in [126].

A symmetric b.i.b.d. with $v = 43$, $r = 7$, $\lambda = 1$ is impossible, because the equivalent $PG(2,6)$ does not exist. It follows that a b.i.b.d. with $v = 36$, $b = 42$, $r = 7$, $k = 6$ cannot exist either, because it would be equivalent to a $EG(2,6)$, which could be extended into a $PG(2,6)$.

For $k = 5$ and $\lambda = 2$ the integrality of the expressions for r and b is not sufficient, as shown in [74]. On the other hand, considering special values of v, it is shown in [56] that one b.i.b.d. exists (unique apart from isomorphisms) for $v = b = 11$, $r = k = 5$, $\lambda = 2$ and that three non-isomorphic b.i.b.d's exist for $v = b = 16$, $r = k = 6$, $\lambda = 2$.

Designs with $v = b = 37$, $r = k = 9$, $\lambda = 2$ are mentioned in [29] and they are studied in [59].

Nandi has shown in [75] that there exists only one design with $v = 6$, $b = 10$, $r = 5$, $k = 3$, $\lambda = 2$. It appeared first in [131] and it is probably the very first b.i.b.d. to have been published. We quote it here:

$$012 \quad 013 \quad 024 \quad 035 \quad 045$$
$$125 \quad 134 \quad 145 \quad 234 \quad 235 \ .$$

Nandi has also shown that there are three designs for $v = 10$, $b = 15$, $r = 6$, $k = 4$, $\lambda = 2$ and in [76] that for any v smaller than 15 there exists only one symmetric b.i.b.d., if at all, while there are five for $v = b = 15$, $r = k = 7$, $\lambda = 3$. From the latter, four non-isomorphic b.i.b.d's can be formed by block section, and four by block intersection.

We shall now derive existence theorems by making use of the concept of the incidence matrix.

If the b.i.b.d. is symmetric, then, as we have seen, $|AA'| = |A|^2 = r^2(r - \lambda)^{v-1}$. Therefore we have

Theorem II.1 If $v = b$ is even, then $r - \lambda = k - \lambda$ must be a square. (This has also been pointed out in [112], [33], and [105].)

Thus, for instance, b.i.b.d's with $v = b = 22$, $k = r = 7$, $\lambda = 2$ or with $v = b = 46$, $k = r = 10$, $\lambda = 2$ are not possible. The former fact is also proved in [57].

For odd v, a necessary condition for a symmetric b.i.b.d. to exist is given by Chowla and Ryser in [33]. To prepare us for its proof we show first

Theorem II.2 If v, k, and λ are parameters of a symmetric b.i.b.d., and v is odd, then the equation $x^2 = (k-\lambda)y^2 + (-1)^{\frac{1}{2}(v-1)}\lambda z^2$ has an integer solution where x, y, and z are not all 0.

Proof: Let the incidence matrix be $A = (a_{ij})$ and consider the expressions $L_j = \sum_{i=1}^{v} a_{ij} x_i$. By the definition of k and λ we have

$$\sum_{j=1}^{v} L_j^2 = (k-\lambda) \sum_{i=1}^{v} x_i^2 + \lambda \left(\sum_{i=1}^{v} x_i \right)^2 \quad . \quad . \quad . \quad (L)^*$$

We refer now to Lagrange's theorem which states that every positive integer is the sum of (not more than) four squares.

Let $k - \lambda = a_1^2 + a_2^2 + a_3^2 + a_4^2$. Then

$$(k-\lambda) = (x_1^2 + x_2^2 + x_3^2 + x_4^2) = (y_1^2 + y_2^2 + y_3^2 + y_4^2)$$

where

$$y_1 = a_1 x_1 + a_2 x_2 + a_3 x_3 + a_4 x_4 \quad y_2 = a_2 x_1 - a_1 x_2 + a_4 x_3 - a_3 x_4$$
$$y_3 = -a_3 x_1 + a_4 x_2 + a_1 x_3 - a_2 x_4 \quad y_4 = a_4 x_1 + a_3 x_2 - a_2 x_3 - a_1 x_4$$

The matrix which transforms the $x - s$ into the $y - s$ is not singular. Its determinant is, in fact, $(k-\lambda)^2$. Thus the L_i are linear rational functions of the y_j, say $L_i = \sum_{j=1}^{v} b_{ij} y_j$.

Conversely, the x_i are linear rational functions of the y_j. The equation (L) is therefore an identity in the y_j and it holds whichever relations we establish between them.

At this stage we have to distinguish between the two cases (i) $v \equiv 1 \pmod 4$ and (ii) $v \equiv 3 \pmod 4$.

---

* Simple arithmetic shows that if $k - \lambda$ is a square, say m^2, then

$$L_j = mx_j + \Sigma (k-m) x_i / v$$

satisfies this equation. The coefficients of the x_i are rational, but to obtain an incidence matrix they would have to be 0 or 1.

Case (i)

$v - 1$ is divisible by 4, so that we can write

$$\sum_{j=1}^{v} L_j^2 = (y_1^2 + y_2^2 + y_3^2 + y_4^2) + \ldots + (y_{v-4}^2 + y_{v-3}^2 + y_{v-2}^2 + y_{v-1}^2)$$

$$+ (k - \lambda) y_v^2 + \lambda \left(\sum_{i=1}^{v} x_i \right)^2$$

where the y_j are linear rational functions of the x_i (and $x_v = y_v$).

If $b_{11} \neq 1$, then let y_1 be such a rational function of $y_2, \ldots, y_v$ that $L_1 = y_1$; if $b_{11} = 1$, such that $L_1 = -y_1$. In either case $L_1^2 = y_1^2$.

We continue treating $L_2, \ldots$, in the same way, thereby making y_2 a rational function of $y_3, \ldots, y_v$ and so on, until $L_1 = y_1, \ldots, L_{v-1} = y_{v-1}$ have cancelled out and we reach

$$L_v^2 = (k - \lambda) y_v^2 + \lambda \left(\sum_{i=1}^{v} x_i \right)^2.$$

Now both sides are rational functions of y_v and we choose it to be different from 0. It follows that $x^2 = (k - \lambda) y^2 + \lambda z^2$ can be solved in rational, and hence in integral values, with $y \neq 0$.

Case (ii)

$v + 1$ is divisible by 4, and introducing $y_{v+1} = x_{v+1}$ we can write

$$\sum_{j=1}^{v} L_j^2 + (k - \lambda) x_{v+1}^2 = (k - \lambda) \sum_{i=1}^{v+1} x_i^2 + \lambda \left(\sum_{i=1}^{v} x_i \right)^2 =$$

$$(y_1^2 + y_2^2 + y_3^2 + y_4^2) + \cdots + (y_{v-2}^2 + y_{v-1}^2 + y_v^2 + y_{v+1}^2) + \lambda \left(\sum_{i=1}^{v} x_i \right)^2.$$

Proceeding as we did in case (i), we obtain eventually

$$(k - \lambda) y_{v+1}^2 = y_{v+1}^2 + \lambda \left(\sum_{i=1}^{v} x_i \right)^2.$$

Hence $x^2 = (k - \lambda) y^2 - \lambda z^2$ can be solved in integers, not all of them zero.

This completes the proof of Theorem II.2.

Examples

(1) Consider the case of a finite projective plane. Then $v = 1 + (k - 1) + (k - 1)^2$ is odd for any k. The Diophantine equation is in case (i), i.e. when $v \equiv 1 \pmod{4}$, i.e. $k - 1 \equiv 0$ or $3 \pmod{4}$, $x^2 = (k - 1) y^2 + z^2$ and can be solved when $k - 1 = 4t$ by $x = 2t + 2$, $y = 2$,

$z = 2t - 2$ and when $k - 1 = 4t + 3$ by $x = 2t + 2$, $y = 1$, $z = 2t + 1$.

In case (ii), i.e. when $v \equiv 3 \pmod 4$, i.e. $k - 1 \equiv 1$ or $2 \pmod 4$ it is $x^2 + z^2 = (k-1)y^2$.

This can only be solved in integers if $k - 1$ is the sum of two squares, which excludes the values $k = 7, 15, 22, 23$, etc. It is not known whether any other values, not belonging to this series, are also excluded from leading to a finite projective plane. On the other hand, it is easily seen that the rule does not exclude any prime powers: $2^n \equiv 2 \pmod 4$ only if $n = 1$; any prime of the form $4t + 1$ is the sum of two squares; $(a^2 + b^2)^{2t}$ is a square, and

$$(a^2 + b^2)^{2t+1} = [a(a^2 + b^2)^t]^2 + [b(a^2 + b^2)^t]^2.$$

(2) For the two Hadamard designs the resulting Diophantine equations are $x^2 = ty^2 - (t-1)z^2$ and $x^2 = ty^2 - tz^2$. The first is solved by $x = y = z = 1$, and the second by $x = 0$, $y = z = 1$.

We prove now the theorem of Chowla and Ryser, namely

Theorem II.3 Let a symmetric b.i.b.d. with parameters v, k, λ with odd v be given. If a prime p divides the square free part of $k - \lambda$, then $(-1)^{\frac{1}{2}(v-1)}\lambda$ is a quadratic residue of p.

Proof: If the assumption of the theorem holds, then the Diophantine equation of Theorem II.2 can be written

$$x^2 = pty^2 + (-1)^{\frac{1}{2}(v-1)}\lambda z^2,$$

where t is square free and relatively prime to p, and where y is the previous y multiplied by the square factor (if any) of $k - \lambda$. We can choose x, y, and z so that they do not have a common factor. Then p does not divide z, because if it did, then it would also divide x, p^2 would divide x^2, and hence pty^2. But then p would have to divide y as well.

Because p does not divide z, $(-1)^{\frac{1}{2}(v-1)}\lambda$ is a quadratic residue of p.

Example

$v = b = 29$, $r = k = 8$, $\lambda = 2$. Here 3 divides 6, and $(-1)^{\frac{1}{2}(v-1)}. 2 = 2$ is not a quadratic residue of 3. Therefore a b.i.b.d. with the given parameters does not exist (cf. also [55]).

Note that the theorem is trivial for $p = 2$, since then both 0 and 1 are quadratic residues.

By methods outside the scope of this book it is proved in [26] that if $s \equiv 1$ or 2 (mod 4), and if the square-free part of s contains at least one prime factor of the form $4t+3$, then no b.i.b.d. with $v = b = 1+s+s^2$, $r = k = 1+s$, $\lambda = 1$ (i.e. a finite projective plane $P(2,s)$), can exist.

The following similar theorem refers to resolvable configurations.

Theorem II.4 If in an affine resolvable b.i.b.d. with parameters v, b, r, k, λ the number v of varieties is odd, then if r is odd, k must be a square, and if r is even, v must be. (In the latter case, since k^2/v is an integer, this is also a square.)

Proof (cf. [63]): Adjoin to the incidence matrix $r - 1$ rows to obtain a square matrix, $\overline{A}$ say. Let the ith new row have 1 in positions $(i-1)n+1$, $(i-1)n+2, \ldots, (i-1)n+n$, and 0 in all other positions. Then the determinant $|\overline{A}\overline{A}'|$ equals

$$\begin{vmatrix} r & \lambda & \ldots & \lambda & 1 & 1 & \ldots & 1 \\ \lambda & r & \ldots & \lambda & 1 & 1 & \ldots & 1 \\ & & - & - & - & - & & \\ \lambda & \lambda & \ldots & r & 1 & 1 & \ldots & 1 \\ 1 & 1 & \ldots & 1 & n & 0 & \ldots & 0 \\ 1 & 1 & \ldots & 1 & 0 & n & \ldots & 0 \\ & & - & - & - & - & & \\ 1 & 1 & \ldots & 1 & 0 & 0 & \ldots & n \end{vmatrix}$$

and can be computed by applying the following steps:

(i) add the sum of the 2nd to vth row to the first (and remember that $r+\lambda(v-1) = rk$);

(ii) subtract the first column from the 2nd, third, $\ldots$, vth (and remember that $r-\lambda = k$);

(iii) subtract $1/k$ times the sum of the 2nd to vth column from the last $r-1$ columns;

(iv) subtract $v/n = k$ times the sum of the last $r-1$ rows from the first row.

We obtain a determinant with zeros in the triangle above the diagonal, and its value is

$$k^{v-1}n^{r-1}(rk - k(r-1)) = k^v n^{r-1} = k^{v-r+1}v^{r-1}$$

This determinant must be a perfect square, and the theorem follows. An equivalent result is proved in [114], using results from [36].

Structure

We turn to the investigation of the structure of b.i.b.d's, in particular as regards the dependence of sets of blocks on others.

Order the blocks of a b.i.b.d. with incidence matrix A in some way and consider only the first t of them. In the incidence matrix, the first t columns refer to these blocks, and we call the incidence matrix consisting of these t blocks A_0.

Consider the matrix $\begin{pmatrix} A_0 & 0 \\ I_t & 0 \end{pmatrix} = A_1$, say, where 0 is the t by $b-t$ zero matrix. We have

$$A_1 A_1' = \begin{pmatrix} AA' & A_0 \\ A_0' & I_t \end{pmatrix}$$

In [36] W.S. Connor has shown by simple arithmetic that the determinant $|A_1 A_1'|$ equals $kr^{1-t}(r-\lambda)^{v-t-1}|C_t|$. Here C_t is called the *characteristic matrix* of the chosen t blocks, and can be written

$$\lambda k J_t + r(r-\lambda)I_t - rA_0'A_0.$$

Because $|A_1 A_1'| \geqslant 0$ if the number of columns is not smaller than that of the rows of A_1, and because it is $= 0$ otherwise, we see that $|C_t| \geqslant 0$ if $b \geqslant t+v$, $|C_t| = 0$ if $b < t+v$ while, if $b = t+v$, $|A_1 A_1'|$ is a perfect square, since all elements of A_1 are integers.

When $t = 1$, then $|C_1| = \lambda k + r(r-\lambda) - rk = (r-\lambda)(r-k) \geqslant 0$, from which we obtain again the inequality $r \geqslant k$ or $b \geqslant v$.

When $t = 2$, then $|C_2| = (r-\lambda)^2(r-k)^2 - (\lambda k - \mu_{12}r)^2 \geqslant 0$. For a symmetric b.i.b.d. this results in $\mu_{12} = \lambda$, also a known result. Another example is given by $v = 16$, $b = 24$, $k = 6$, $r = 9$, $\lambda = 3$ which leads to $0 \leqslant \mu \leqslant 4$.

Connor has studied in particular the b.i.b.d's with $v = \binom{k+1}{2}$, $b = \binom{k+2}{2}$, $r = k+2$, $\lambda = 2$ (the case for $k = 3$ is quoted on page 15).

Take a block B_1, and call n_i the number of those other blocks which have i varieties in common with B_1. Then

$$\sum_{i=0}^{k} n_i = b-1 = \tfrac{1}{2}k(k+3), \qquad \sum_{i=0}^{k} i n_i = k(r-1) = k(k+1)$$

$$\sum_{i=0}^{k} \binom{i}{2} n_i = \binom{k}{2}$$

It follows that

$$\sum_{i=0}^{k} (i-1)(i-2) n_i = \sum_{i=0}^{k} i(i-1) n_i - 2 \sum_{i=0}^{k} i n_i + 2 \sum_{i=0}^{k} n_i$$

$$= k(k-1) - 2k(k+1) + k(k+3) = 0.$$

Because i as well as n_i is non-negative, all terms $(i-1)(i-2)n_i = 0$, since otherwise their sum could not be zero. Hence $n_i = 0$ for all i except for $i = 1$, or $i = 2$. For these two values we have, from $n_1 + n_2 = \tfrac{1}{2}k(k+3)$ and $n_1 + 2n_2 = k(k+1)$ that $n_1 = 2k$, and $n_2 = \tfrac{1}{2}k(k-1)$. Thus the matrix $A'A$ has k in all diagonal positions, and either 1 or 2 in all other positions.

By a similar arithmetic Connor establishes, starting from two initial blocks, that if these have $t(= 1$ or $2)$ varieties in common, then there exist

$(k-t)(k-t+1) + kt - \tfrac{1}{2}k(k+3) + 1$ blocks with two varieties from each of the initial blocks,

$2t(k-t)$ blocks with two varieties from one of them and one variety from the other, and

$k(2-t) - 2t(1-t)$ blocks with one variety from each of the two initial blocks.

EXERCISES

(1) Construct a resolvable b.i.b.d. from the conic in the $PG(2,4)$ given in the example on page 4 by the method explained on page 10.

(2) Construct a design which is complementary to the $PG(2,2)$.

(3) Construct a Hadamard matrix from the $PG(2,2)$.

(4) Illustrate Theorem II.2 in the case of a $PG(2,2)$.

CHAPTER III

LATIN SQUARES AND ORTHOGONAL ARRAYS

Latin squares

Consider n^2 cells, laid out in a square pattern, and containing n different symbols, each repeated n times, in such a way that the cells in each row, and also in each column, contain each symbol precisely once. Such an arrangement is called a Latin square (L.Sq.), because L. Euler, who studied such patterns in [40], used Latin letters for symbols. We call n the side of the L.Sq.

An obvious way of constructing a L.Sq. is to write down the n symbols in some order and then to permute them cyclically to obtain other rows. A L.Sq. with side 5, constructed by moving each line two positions on to obtain the next, is attributed to the Norwegian Knut Vik in [44]. It is also easily seen that the elements of a group, written in the form of their multiplication table, form a L.Sq.

Starting from a L.Sq. and permuting symbols, rows, or columns, or by reflections, we obtain in general new L.Sq's. However, we do not always obtain, by these devices, all L.Sq's of a given side. We shall not deal here with questions of enumeration, but mention a few pertinent points. The symbols can be chosen in such a way that they appear in a prescribed order in the first row, and so that they appear in the first column in the same order. We say, then, that the L.Sq. is in Standard Form, and $n! \cdot (n-1)!$ L.Sq's are represented by the same Standard Form.

There exists just one L.Sq. in Standard Form for $n = 2$, one for $n = 3$, and four for $n = 4$, namely:

0123	0123	0123	0123
1230	1032	1302	1032
2301	2310	2031	2301
3012	3201	2031	3210
		3210	

For higher n the numbers increase rapidly. They are: 56 for $n = 5$ (all found by Euler), 9408 for $n = 6$ (see [45]) and 16,942,080 for $n = 7$ (see [77] and [104]).

Further information can be found in [77] and in Chapter 4 of [48]. Interesting facts are also given in [98].

Latin rectangles

A Latin rectangle of c columns and $n > c$ rows is an array of n symbols such that each column contains all the symbols, and no row contains any symbol more than once. (This pattern is also called a *Youden square*.) If we take the rows as blocks, then we obtain a symmetric configuration. The complementary design can also be written as a Youden square, as was proved in [129], and the two designs can be combined into a L.Sq. In other words, any Latin rectangle can be extended into a L.Sq. (cf. [47]). For the number of ways in which this can be done, see [39].

We shall now show that we can also extend an r by c rectangle, containing n symbols in such a way that no symbol appears twice in any row or in any column, into a L.Sq. by adjoining $n - c$ columns and $n - r$ rows, provided the number $N(i)$ of times that the symbol i appears in the rectangle is at least $r + c - n$, for all symbols.

The necessity of the condition is obvious. If we want to extend the given rectangle into a Latin rectangle of n rows and c columns (as we must, at least implicitly) then we shall have, in the latter, c repetitions of every symbol. But in the additional $n - r$ rows we cannot include any symbol more than $n - r$ times, hence it must already appear $c + r - n$ times in the original pattern.

To prove the sufficiency, form the sets S_j, consisting of those elements which are not included in the jth column $(j = 1, \ldots, c)$ of the original pattern. Each of these sets consists of $n - r$ elements, and each symbol will, altogether, appear $M(i) = c - N(i)$ times. To write down these $S_j (j = 1, \ldots, c)$ as complementary to the c columns in the original pattern, we must be able to choose a system of c distinct elements, one from each set, which we then use as a further row. Such a system exists by P. Hall's theorem (see [129]), because any k sets of S_j contain altogether $k(n - r)$ elements, none more often than $n - r$ times, in view of $M(i) = c - N(i) \leqslant c - (c + r - n)$ for all i. Continuing, we can thus construct $n - r$ additional rows to complete an n by c Latin rectangle, and from this we can complete a Latin square, as we have seen.

This proof is contained, in a slightly different form, in [103].

Example

$n = 6$, $c = 3$, $r = 4$

123	i	$M(i)$	$N(i)$	i	$M(i)$	$N(i)$
246	1	2	1	4	1	2
352	2	0	3	5	1	2
465	3	1	2	6	1	2

missing 511 which can be written 531
  634   614

The Latin rectangle 123 can be completed by adding 456
                     246   (for instance)   315
                     352   641
                     465   132
                     531   264
                     614   523

As a consequence of the results just derived it is shown in [41] that an incomplete L.Sq. of side $n (\geqslant 4)$ with t different elements (i.e. a square of side n such that a subset of the n^2 places are occupied by numbers from among 1, 2, ..., n, and no element occurs twice in the same row or column) can be embedded in a L.Sq. of side t, if $t \geqslant 2n$.

Orthogonal sets of Latin squares

In what follows, we shall be interested in pairs of so-called orthogonal Latin squares, i.e. L.Sq's of equal sides and such that their superposition produces ordered pairs of elements, each of which occurs only once in the same cell. For example, the two L.Sq's of side 3,

$$\begin{array}{ccc} 012 & & 012 \\ 120 & & 201 \\ 201 & & 120 \end{array}$$

are orthogonal. Such a pair is also called a *Graeco-Latin square*, following Euler who wrote the symbols of the first in Latin and those of the second in Greek letters, thus:

$$\begin{array}{ccc} a\alpha & b\beta & c\gamma \\ b\gamma & c\alpha & a\beta \\ c\beta & a\gamma & b\alpha \end{array}$$

A set of L.Sq's such that any two of them are orthogonal is called an *orthogonal set*. Some writers use such expressions as "mutually", or "pairwise orthogonal" L.Sq's. Others, imagining all of them superimposed, speak of "hyper-Graeco-Latin" squares.

There cannot be more than $n-1$ orthogonal L.Sq's in any set. This is seen as follows.

The symbols in each square can be permuted, independently, without destroying orthogonality. The first row in all the squares can therefore be written as $0\ 1\ .\ .\ .\ n-1$.

The symbol 1 will in any pair correspond to itself in the top row. In the first column, this symbol cannot occupy the top position in any of the squares, because this position is already occupied by 0. Furthermore, it cannot occupy any other position in that column more than once taking all the squares together, because it already corresponds to itself in the first row. Hence, not more than $n-1$ squares can be constructed to form an orthogonal set. If a set contains precisely $n-1$ orthogonal L.Sq's then it is called *complete*. (We shall see in Chapter V that the existence of $n-3$ pairwise orthogonal L.Sq's of order $n>4$ implies the existence of a complete set.)

When n equals a prime power, s say, then a complete set can be constructed from a finite projective plane.

Choose a particular line L. Through each of its $s+1$ points there pass s lines, apart from L. We call these a *pencil* of lines, and in each of the $s+1$ pencils we number the lines $0, 1, ..., s-1$. Through each of the s^2 points outside L there passes one line from each of the $s+1$ pencils.

Denote the points on L by $P_0, P_1, ..., P_s$. Select two of these, say P_0 and P_1, and let the numbers of the lines through them refer, respectively, to rows and columns of a square arrangement. Enter in the cell of row i and column j the number of that line from P_2 which passes through the point defined by the line i from P_0 and line j from P_1. Because any two lines meet in one single point, we obtain in this way a L.Sq.

Construct other L.Sq's by using P_3 instead of P_2, and so on. It is easily seen that – again because two lines meet in a single point – any two of the $s-1$ L.Sq's are orthogonal.

Conversely, if $n-1$ pairwise orthogonal L.Sq's exist, then a finite projective plane with $n+1$ points on each line also exists. To prove this, it is sufficient to show that a b.i.b.d. with $v=n^2$, $b=n(n+1)$, $r=n+1$, $k=n$, $\lambda=1$ exists, i.e. a $EG(2,n)$ which can be expanded into a projective plane.

For the purpose of this proof, number the cells of the L.Sq's from 1 to n in the first row, $n+1$ to $2n$ in the second, etc., until n^2-n+1 to n^2 in the last row. Then form blocks, by putting into the same block the numbers of those cells which have the same element in the 1st, 2nd, ..., $(n-1)$th L.Sq. This gives $n(n-1)$ blocks of n varieties each. To these add the $2n$ blocks which form the rows and the columns of the square with the numbers from 1 to n^2.

It is not known whether a complete set of orthogonal L.Sq's exists when the side n is not a prime power.

If we start from a $PG(2,s)$, then we can take as the first two pencils the lines $x_1 = a_1$, and $x_2 = a_2$, where a_1 and a_2 take all the values of $GF(s)$, and number the rows by a_1 and the columns by a_2. In the cell (a_1, a_2) enter the value $a_1 x + a_2$, where x is a given non-zero element of $GF(s)$. The various orthogonal L.Sq's of the complete set are obtained by letting x have the various non-zero values. (This "Bose-Stevens" construction was given, independently, in [125] and in [3].)

If n is a prime, and x its primitive root, then

$$
\begin{array}{cccc}
0 & 1 & \ldots & n-1 \\
x^i & x^i + 1 & \ldots & x^i + n - 1 \\
x^{i+1} & x^{i+1} + 1 & \ldots & x^{i+1} + n - 1 \\
& & - \ - \ - & \\
x^{i+n-2} & x^{i+n-2} + 1 & \ldots & x^{i+n-2} + n - 1
\end{array}
$$

is a L.Sq., and two L.Sq's for different i are orthogonal. The proof is immediate. In this case the various L.Sq's for the different values of $i = 0, 1, \ldots, n-2$ are obtained by cyclic permutation of the last $n-1$ rows.

Example

Let $s = 3$. Then we have, for

$$
\begin{array}{cc}
x = 1 & x = 2 \\
\begin{array}{ccc} 0 & 1 & 2 \\ 1 & 2 & 0 \\ 2 & 0 & 1 \end{array} &
\begin{array}{ccc} 0 & 1 & 2 \\ 2 & 0 & 1 \\ 1 & 2 & 0 \end{array}
\end{array}
$$

Conversely, if we number the cells

$$1 \quad 2 \quad 3$$
$$4 \quad 5 \quad 6$$
$$7 \quad 8 \quad 9$$

then we obtain a b.i.b.d. (which is resolvable)

123	147	168	159
456	258	249	267
789	369	357	348

In such a complete set any two L.Sq's differ only through a permutation of their rows (this is called "property D_0" in [17]). [46] contains a complete set of L.Sq's of side 9, with property D_0, which does not arise from a $PG(2, 9)$.

A collection of n cells in a L.Sq. of side n, such that no row and no column contains two cells of the collection, and no two cells of it contain the same symbol, is called a *transversal* of the L.Sq. If the same collection of cells is a transversal in every L.Sq. of a set, then we call it a transversal of the set.

For instance, if in an orthogonal set we single out one of the L.Sq's and consider the cells which contain the same symbol, then these cells form a transversal of the set of the other L.Sq's.

We now show that a complete set of orthogonal L.Sq's cannot have a transversal. Start with a cell in the first row. We may assume that the first row is the same in all L.Sq's and that the symbol we start with is 0.

To continue with the transversal, $n-1$ cells of the second row are available. Take any one of these cells. It must not contain 0, and its contents must be different in the $n-1$ orthogonal L.Sq's. Also, it must not be the same as the symbol above it in the first row. Clearly, this is impossible. Thus there cannot even be two cells which satisfy the conditions for a transversal of a complete set of orthogonal L.Sq's.

By an application of this result we now prove

Theorem III.1 If $r-1$ L.Sq's of side n form an orthogonal set, and if each L.Sq. contains a subsquare of side r such that all these subsquares form a complete orthogonal set between them, then it is impossible to add an rth L.Sq. of side n orthogonal to all the original ones, unless $n \geqslant r^2$. Moreover, if n is larger than r^2, then it must be at least r^2+r (cf. [85]).

Proof: Without loss of generality, the subsquares may be assumed to be all in the same position. If another L.Sq., orthogonal to all the $r-1$

given ones, can be added, then the set of the $r-1$ L.Sq's has n transversals, namely the collections of cells which contain all 0, or all 1, etc. in the added L.Sq. These transversals do not overlap. As we have just seen, no two cells of a subsquare can belong to the same transversal, because the $r-1$ subsquares form a complete orthogonal set; on the other hand, every cell of a subsquare lies on some transversal. Therefore, the number of transversals must be at least as large as the number of cells in a subsquare, i.e. $n \geqslant r^2$.

If n exceeds r^2, then there must exist a transversal with no cell in a subsquare. Consider the two rectangles within the L.Sq. and outside the subsquare which have, respectively, r rows or r columns in common with the subsquare. Each contains r cells of a transversal, and therefore the remaining $n-2r$ cells of the transversal must be in the subsquare of side $n-r$, complementary to the original subsquare.

The elements $x_1, ..., x_r$ of the subsquare must appear in those $n-2r$ cells. But if a cell in one of the L.Sq's contains x_1, then the corresponding cell in any other L.Sq. cannot contain the same symbol, because the pair (x_1, x_1) is already accounted for within two orthogonal subsquares of side r. Therefore, $r-1$ different cells of the transversal must be available to contain x_1 in the complementary sub-spaces of the several L.Sq's. The same applies to the other x_i, and hence $n-2r \geqslant r(r-1)$, i.e. $n \geqslant r^2 + r$.

Two slightly more general theorems are also proved in [85].

Incidentally, Parker asserts that there *do* exist Graeco-Latin squares of side 10 without Graeco-Latin subsquares of side 3, and such squares were generated on an automatic computer (cf. [82]).

Graeco-Latin squares based on groups

Every row of a L.Sq. contains the same symbols, so that the second, third, etc. rows are permutations of the symbols of the first. Denote the permutation which carries the symbols of the first row into those of the ith by P_i. If $i \neq j$, then $P_i P_j^{-1}$ does not leave any symbol unchanged, because if it did, then that symbol would be in the same column after permutation by P_i and also by P_j. This property of $P_i P_j^{-1}$ for all pairs of different i and j is also sufficient for a L.Sq. to be generated. The P_i do not necessarily form a group, of course.

Theorem III.2   Let the permutations $P_{11}, ..., P_{1n}$ generate a L.Sq. of side n, and let $P_{21}, ..., P_{2n}$ generate another, orthogonal to the first. Then the permutations $P_{11}^{-1} P_{21}, ..., P_{1n}^{-1} P_{2n}$ will also generate a L.Sq.

Proof: $P_{1i}^{-1}P_{2i}$ transforms the ith row of the first into the ith row of the second, and since each pair occurs precisely once in a Graeco-Latin square, the set $P_{1i}^{-1}P_{2i}$ contains, for every pair x, y, a permutation which transforms x into y. There can only be one such transformation.

The converse of Theorem III.2 is also true: if P_{1i} ($i = 1, ..., n$) generate a L.Sq., and $P_{1i}^{-1}P_{2i}$ ($i = 1, ..., n$) generate another, then so do the P_{2i}, and the square generated by the latter is orthogonal to the first, because otherwise $P_{1i}^{-1}P_{2i}$ would not generate a L.Sq.

It follows by induction that the L.Sq's generated by the sets P_{1i}, ..., P_{ri} form an orthogonal set if and only if there exist $r(r-1)$ L.Sq's generated by $P_{ji}^{-1}P_{ki}$ ($j, k = 1, ..., r; j \neq k$).

If the rows in two orthogonal L.Sq's are the same, though in different order, and if the permutations of the rows form a group, then we say that the Graeco-Latin square is based on that group.

For instance, any two orthogonal L.Sq's constructed by the method of Bose and Stevens (page 26), are based on a group. Indeed, let P_1 be the permutation of $a_{10}x + a_2$ into $a_{11}x + a_2$, and P_2 that of $a_{10}x + a_2$ into $a_{12}x + a_2$, then $P_1 P_2$ permutes the row of symbols $a_{10}x + a_2$ into the row $(a_{11} + a_{12} - a_{10})x + a_2$. Hence the permutations form a group. Also, two orthogonal L.Sq's $a_1x_1 + a_2$ and $a_1x_2 + a_2$ contain the same rows in different order. Clearly the rows $a_{11}x_1 + a_2$ and $a_{12}x_2 + a_2$ are the same, if $a_{11}x_1 = a_{12}x_2$.

If two L.Sq's are based on a group, and if their permutations are, respectively, P_i and Q_i ($i = 1, ..., n$), then $P_i^{-1}Q_i = R_i$, say, are also permutations of the same group, and they define a L.Sq. The mapping of the P_i on to the R_i is a complete mapping. Thus, if a group has no complete mapping (e.g. a group of order $4t + 2$), then a Graeco-Latin square of that side (may exist, but) cannot be based on that group.

Conversely, if a complete mapping of a group does exist, then a Graeco-Latin square based on that group can be constructed. Represent the group by a regular permutation group P_i, mapped on R_i (elements of the same regular permutation group). Then, since such a group contains no elements, apart from the identity, which leaves any symbol unchanged, P_i, R_i, and $P_i R_i$ will all generate L.Sq's, and the first and the third are orthogonal. This follows from the converse of Theorem III.2, because $(P_i^{-1})(P_i R_i) = R_i$.

Example

P_i	R_i	$P_i R_i$
I	I	I
(12) (34)	(13) (24)	(14) (23)
(13) (24)	(14) (23)	(12) (34)
(14) (23)	(12) (34)	(13) (24)

The L.Sq's

1234	1234
2143	4321
3412	2143
4321	3412

are orthogonal. A third L.Sq., orthogonal to these two, arises when we interchange the roles of R_i and $P_i R_i$. This is here possible, because $P_i \cdot P_i R_i = R_i$.

The L.Sq.

1234

3412   completes the set.

4321

2143

Existence theorems

We shall now mainly be interested in the existence of Graeco-Latin squares, i.e. of pairs of orthogonal L.Sq's, and start with some preliminary considerations.

There are L.Sq's which cannot possibly be one of an orthogonal pair. An example is given by the following–

Theorem III.3 (see [66]) If in a L.Sq. of side $4t+2$ (t being a positive integer) not more than t cells of the first $2t+1$ rows and columns contain symbols from a given list of $2t+1$ symbols, then the L.Sq. cannot be part of a Graeco-Latin square.

Proof: Divide the L.sq. into four squares, each of side $2t+1$, and denote them as follows:

I   II

III   IV

If any symbol occurs in I m times, then it will occur $2t+1-m$ times in II as well as in III, and hence again m times in IV. Thus in I and IV together (briefly, in I/IV) each of the $4t+2$ symbols appears an even number of times.

Assume now that there exist two orthogonal L.Sq's L and L'. Let the symbols in L be 0, 1, ..., $4t+1$ and call the symbols 0, ..., $2t$ A-marks, and the others B-marks. Assume that in I of L there are not more than t cells with a B-mark in them. Then there will not be more than $2t$ of the A-marks outside I/IV and therefore at least $2t+2$ symbols of L' occur, within I/IV, together with all the A-marks within I/IV.

However, at most $2t$ symbols of L' occur in I/IV together with B-marks, and hence at least 2 remaining symbols of L' (out of those $2t+2$) appear, in I/IV, only with A-marks and therefore an odd number of times. This contradicts the fact, mentioned above, that in any L.Sq. every symbol appears an even number of times within I/IV. No orthogonal pair can therefore exist under the circumstances.

It is clear that we may permute rows, columns, and symbols of L and obtain an equivalent theorem. Hence a L.Sq. of side $4t+2$ which contains a L.Sq. of side $2t+1$ cannot be one of a pair of orthogonal squares. For instance, a multiplication table of a group of order $4t+2$ forms a L.Sq. and contains one of side $2t+1$, which is the multiplication table of a subgroup.

Also, if we construct a L.Sq. of side 6 from 012345 by cyclic permutations, then this L.Sq. contains L.Sq's of side 3, consisting of alternate rows and columns. Tarry has, in fact, proved ([126]) that no Graeco-Latin square of side 6 exists and, as Mann [66] has pointed out, the above approach could simplify his proof.

Ostrowski and Duren have proved ([80]) that the result in Theorem III.3 is best, in the sense that if more than t cells of I contain B-marks, a L.Sq. orthogonal to that given may exist. They exhibit the following examples of side 10:

01234	56789	01923	84657
34012	79865	67895	23104
43120	97658	93746	58210
12407	85396	38254	79061
20375	68941	14507	36982
57698	34120	25619	40873
89756	12034	40138	62795
65981	43207	56480	17329
98563	01472	82071	95436
76849	20513	79362	01548

Three cells in the upper left-hand quarter of the first L.Sq. contain

B-marks, and yet the second L.Sq. is orthogonal to the first.

Mann has also proved in [66] that if in a L.Sq. of side $4t+1$ the square formed by the first $2t$ rows and columns contains fewer than $t/2$ symbols out of a list of $2t$, then this L.Sq. cannot be a member of an orthogonal pair.

Orthogonal arrays

For our further investigations we write sets of orthogonal L.Sq's in a more convenient way. If the side is **n**, then we write two rows of n^2 elements each, thus:

$$0\ 0\ ...\quad 0\quad 1\ 1\ ...\quad 1\quad ...\quad (n-1)(n-1)...(n-1)$$
$$0\ 1\ ...\ (n-1)\ 0\ 1\ ...\ (n-1)\quad ...\quad 0\quad 1\ ...(n-1)$$

The same symbols will also be used for the elements of the L.Sq's.

Imagine the rows and the columns numbered from 0 to $n-1$. We then write into the next row, underneath the symbols $\left|\begin{smallmatrix}i\\j\end{smallmatrix}\right|$, the entry of the first L.Sq. in the cell of the ith row and jth column. Each further row is allocated to a further L.Sq. of the set, in the same manner.

Since the L.Sq's are pairwise orthogonal, the array which we obtain has the property that in any two rows all possible combinations $\left|\begin{smallmatrix}i\\j\end{smallmatrix}\right|$ appear precisely once. The array has n^2 columns and there are in it **n** different symbols. If it has **c** rows, corresponding to $c-2$ L.Sq's, then we call it an orthogonal array of size n^2, c constraints, level n, strength 2, and denote it by $[n^2, c, n, 2]$. If the orthogonal set is complete, then $c = n+1$. Such an array is equivalent to a resolvable b.i.b.d. with $v = n^2$, $b = n(n+1)$, $r = n+1$, $k = n$, $\lambda = 1$, which we construct as follows.

Number the columns and the rows of the array. The number of a column will appear in the ith set, in that block which is indicated by the element of that column in the ith row.

Example

Two orthogonal Latin squares of side 3 are equivalent to the orthogonal array

$$\begin{array}{ccccccccc}
0 & 0 & 0 & 1 & 1 & 1 & 2 & 2 & 2 \\
0 & 1 & 2 & 0 & 1 & 2 & 0 & 1 & 2 \\
0 & 1 & 2 & 2 & 0 & 1 & 1 & 2 & 0 \\
0 & 1 & 2 & 1 & 2 & 0 & 2 & 0 & 1
\end{array}$$

This leads to the resolvable b.i.b.d. on page 27.

More general orthogonal arrays are defined as follows.

An orthogonal array of size N, c constraints, level n and strength t, written $[N, c, n, t]$, is a rectangular arrangement of n different symbols in N columns and c rows, and such that in any t rows the N columns contain all the n^t possible ordered sets of n elements the same number of times. If we call this number the *index* of the array, λ, then clearly $N = \lambda n^t$. (These definitions are due to C.R. Rao, see [90].)

If we retain only $c'(\geqslant t)$ rows from such an array, then we have a $[N, c', n, t]$ with the same index. An array of strength t and index λ is also of strength t' and index $\lambda n^{t-t'}$, $(t > t')$. If λ is a power of n, then the array is called a *hyper-Latin cube*. When $\lambda = 1$, and $t = 2$, then such a hyper-Latin cube is the same as an orthogonal set of L.Sq's.

Raghavarao defines in [87] a Latin cube of the first order, of side n, as a cube of n letters n^2 times repeated, each letter being exactly n times in each of its planes (there are three parallel sets of these, with n planes in each set). Two such cubes are orthogonal if every letter of one cube appears n times in the same position with every other letter in the other cube. A set of c pairwise orthogonal cubes is equivalent to a $[n^3, c, n, 2]$. We shall see that the maximum number of cubes in a set of pairwise orthogonal ones is $n^2 + n - 2$ (see page 42). A set consisting of this number of cubes is called *complete*. The construction was studied in [60] and [24].

We give here an example of a complete set of 10 orthogonal Latin cubes of side 3 from [42].

012	012	012	012	000	000	012	021	021	021
120	201	012	012	111	111	120	102	201	120
201	120	012	012	222	222	201	210	120	201
012	012	120	201	111	222	120	102	120	201
120	201	120	201	222	000	201	210	012	012
201	120	120	201	000	111	012	012	201	120
012	012	201	120	222	111	201	210	201	120
120	201	201	120	000	222	012	021	120	201
201	120	201	120	111	000	120	102	012	012

Construction

In order to construct an orthogonal array, we may turn to a $PG(m, s)$, following C.R. Rao ([90]). After omitting a selected

$(m-1)$-flat "at infinity", we obtain a $EG(m, s)$. The $(m-1)$-flat at infinity contains $(s^m-1)/(s-1)$ different $(m-2)$-flats, and through each of them there pass s different $(m-1)$-flats, apart from that at infinity. We call each of the $(m-2)$-flats at infinity a *vertex* of a parallel pencil of $(m-1)$-flats, and there will be one $(m-1)$-flat from each of the $(s^m-1)/(s-1)$ pencils through every point of the $EG(m, s)$. Any two $(m-1)$-flats from different pencils intersect in an $(m-2)$-flat with s^{m-2} points.

To the $(m-1)$-flats of each pencil attach the elements of a $GF(s)$, number the pencils in some way, and attach to each of the s^m points of the $EG(m, s)$ the ordered set of the numbers of the $(m-1)$-flats passing through it. If we give a column to each point and a row to each pencil, then we have s^m columns and $(s^m-1)/(s-1)$ rows. Write into the column of each point the numbers of its $(m-1)$-flats. In any two rows any combination of the elements of the $GF(s)$ appears then s^{m-2} times (this being the number of points in the $(m-2)$-flat in which the two pencils, represented by the rows, intersect), and we have thus obtained a $\left[s^m, \dfrac{s^m-1}{s-1}, s, 2\right]$. It is equivalent to an affine resolvable b.i.b.d. with $v = s^m$, $b = s(s^m-1)/(s-1)$, $k = s^{m-1}$, $r = (s^m-1)/(s-1)$, $\lambda = (s^{m-1}-1)/(s-1)$.

Another construction of arrays of strength 2 is based on Hadamard matrices. Let such a matrix of order 4λ be given. It can always be such that all elements in the first row are 1. Omitting this row, and replacing -1 by 0, we obtain an array $[4\lambda, 4\lambda-1, 2, 2]$. The proof is almost immediate. The resulting arrays are explicitly given in [86] for all $\lambda \leqslant 25$, except for $\lambda = 23$.

Also in [86], Plackett and Burman have constructed orthogonal arrays by cyclically permuting a row of s^m-1 elements $(s^m-1)/(s-1)$ times and adding a column consisting entirely of the element 0.

More general arrays of strength t can be constructed if we can find c pencils in a $EG(m, s)$ such that any t $(m-1)$-flats from different pencils have s^{m-t} points in common, i.e. such that they intersect in an $(m-t)$-flat. This means that the vertices of the pencils (they are again $(m-2)$-flats in a $PG(m, s)$ in which the $EG(m, s)$ is embedded) must be such that any t of them have an $(m-t-1)$-flat, but no flat of higher dimension in common. We then obtain an $[s^m, c, s, t]$.

The $(m-1)$-flats of a pencil have analytic representations $a_0 + a_1 x_1 + \cdots + a_m x_m = 0$ where the $a_i (i = 1, \ldots, m)$ are fixed and the value of a_0 varies. Let a_0 be the value attached to the $(m-1)$-flat

whose equation contains it, out of a pencil defined by the other a_i.
Let c pencils be given by $(a_{j1}, \ldots, a_{jm})$ $(j = 1, \ldots, c)$. Select c $(m-1)$-
flats, one from each pencil, denoting them by a_{j0}. We want any t of
them to intersect in a $(m-t)$-flat, whatever the a_{j0}. Hence any submatrix of t rows of the matrix

$$
\begin{pmatrix}
a_{11} & \cdots & a_{1m} \\
& - - - & \\
a_{c1} & \cdots & a_{cm}
\end{pmatrix}
$$

must have rank t. Each point $(x_1, \ldots, x_m)$ in a $EG(m, s)$ receives
from the jth pencil the mark a_{j0}, given by

$$a_{j0} + a_{j1} x_1 + \ldots + a_{jm} x_m = 0.$$

Therefore the set $(a_{10}, \ldots, a_{t0})$ will be allotted to the solutions of

$$a_{10} + a_{11} x_1 + \ldots + a_{1m} x_m = 0$$

$$- - - - -$$

$$a_{t0} + a_{t1} x_1 + \ldots + a_{tm} x_m = 0,$$

i.e. altogether to s^{m-t} points.

It follows that if a matrix, as defined above, is given, then the s^m
columns

$$a_{11} x_1 + \ldots + a_{1m} x_m$$

$$- - - - -$$

$$a_{c1} x_1 + \ldots + a_{cm} x_m$$

formed from the s^m sets $(x_1, \ldots, x_m)$ will form an array $[s^m, c, s, t]$.
(Cf. [11].)

A matrix of the required property is that of the co-ordinates of c
points in a $PG(m, s)$, where no t points lie in a space of less than
$t-1$ dimensions.

Example (from [11]).
    The ten points of the quadric $x_1^2 + x_2^2 = x_3 x_4$ in a $PG(3, 3)$ are

$$(0010) \quad (0001) \quad (0111) \quad (0122) \quad (1011)$$
$$(1022) \quad (1112) \quad (1121) \quad (1212) \quad (1221)$$

No three of these points are on the same line. Therefore the co-
ordinates can be used as the rows of a matrix, from which an array

[81, 10, 3, 3] can be constructed.

Esther Seiden has given in [106] rules for constructing arrays of strength 3 from suitable ones of strength 2, and also otherwise. We give here a simple example.

Take any three rows of a [4λ, c, 2, 2]. There will be λ columns,

$$\begin{array}{cc} 0 & 1 \\ \end{array} \qquad \begin{array}{cc} 0 & 1 \end{array}$$

each of which is either 0 or 1, λ columns either 0 or 1, λ either

$$\begin{array}{cc} 0 & 1 \\ \end{array} \qquad \begin{array}{cc} 1 & 0 \end{array}$$

$$\begin{array}{cc} 0 & 1 \end{array} \qquad \begin{array}{cc} 1 & 0 \end{array}$$

1 or 0, and λ either 0 or 1.

$$\begin{array}{cc} 0 & 1 \end{array} \qquad \begin{array}{cc} 0 & 1 \end{array}$$

Proof: Let x_{ijk} denote the number of columns $\begin{smallmatrix} i \\ j \\ k \end{smallmatrix}$ in three rows chosen. Then

$$\sum_i x_{ijk} = \sum_j x_{ijk} = \sum_k x_{ijk} = \lambda.$$

For instance, $x_{000} + x_{010} = x_{010} + x_{110} = x_{110} + x_{111} = \lambda$, so that $x_{000} + x_{111} = \lambda$. Similarly, $x_{001} + x_{110} = x_{010} + x_{101} = x_{100} + x_{011} = \lambda$. Thus, if to a given array we add all columns obtained by changing in those already given 0 into 1 and vice versa, then we obtain a [8λ, c, 2, 3], and adding a row consisting of 4λ times 0 and 4λ times 1, we have a [8λ, c+1, 2, 3].

Example

	0011	00111100
[4, 3, 2, 2]	0101 ⟶ [8, 4, 2, 3]	01011010
	0110	01101001
		00001111

Another method of constructing orthogonal arrays will now be explained.

There are s^t different polynomials

$$a_{t-1} x^{t-1} + \ldots + a_1 x + a_0$$

with coefficients from a $GF(s)$. We call them $y_j(x)$, $j = 1, 2, \ldots, s^t$. Denote the elements of the $GF(s)$ by $x_1, \ldots, x_s$. Let $t \leqslant s$. Then the matrix $y_j(x_i)$, i for rows and j for columns, is an orthogonal array $[s^t, s, s, t]$.

For the proof, (from [27]), take any t rows. Without loss of generality, we may assume that they are the first t. Since the array has s^t columns, it is sufficient to show that any two columns of the t rows are different. The identity of two columns would mean that

$$\sum_{j=0}^{t-1} a_j x_i^j \ = \ \sum_{j=0}^{t-1} a_j' x_i^j, \ \text{i.e.} \ \sum_{j=0}^{t-1} (a_j - a_j') x_i^j \ = \ 0$$

for $i = 1, ..., t$. But the Vandermonde determinant

$$\begin{vmatrix} 1 & x_1 & x_1^2 & ... & x_1^{t-1} \\ & & - - - - - & & \\ 1 & x_t & x_t^2 & ... & x_t^{t-1} \end{vmatrix}$$

is only zero if two of the x_i are equal, and since this is not the case here, the only solution of the system of equations in the $(a_j - a_j')$ is $a_j - a_j' = 0$ for all j, so that the two polynomials would then be identical. Thus the s^t different polynomials lead to s^t different columns.

The $[s^t, s, s, t]$ which we have obtained can be expanded into a $[s^t, s+1, s, t]$ by the addition of a row which contains, in each column, the leading coefficient (i.e. a_{t-1}) of the polynomial to which that column corresponds. For the proof we need only investigate those pairs of columns in which the two added elements are equal. Then $a_{t-1} = a_{t-1}'$ and for $\sum_{j=0}^{t-2} a_j x_i^j - \sum_{j=0}^{t-2} a_j' x_i^j = 0$ to hold, a Vandermonde determinant of order $t-1$ would have to be zero. We conclude, as before, that the array is still orthogonal.

Example

$s = 4$, $t = 2$. We obtain a $[16, 5, 4, 2]$.

The elements of $GF(4)$ are $0, 1, y, z = 1+y$. The multiplication table is

	0	1	y	z
0	0	0	0	0
1	0	1	y	z
y	0	y	z	1
z	0	z	1	x

and the array is

0	0	0	0	1	1	1	1	y	y	y	y	z	z	z	z
0	1	y	z	1	0	z	y	y	z	0	1	z	x	1	0
0	y	z	1	1	z	y	0	y	0	1	z	z	1	0	y
0	z	1	y	1	y	0	z	y	1	z	0	z	0	y	1
0	1	y	z	0	1	y	z	0	1	y	z	0	1	y	z

The last line gives the leading coefficients of the respective polynomials, namely

$$0 \quad x \quad yx \quad zx \quad 1 \quad x+1 \quad yx+1 \quad zx+1 \quad y \quad x+y \quad yx+y \quad zx+y$$
$$z \quad x+z \quad yx+z \quad zx+z.$$

If s is a power of 2 and $t = 3$, then yet another row can be added, by writing into each column the coefficient a_{t-2} of the corresponding polynomial. For three rows which do not contain either of the rows of a_{t-1} and a_{t-2}, the argument is the same as before. It may be reduced to that argument if we consider the two rows and a third. However, if only one of the added rows and two of the original ones are taken, then the relevant determinant is either $\begin{vmatrix} 1 & x_1 \\ 1 & x_2 \end{vmatrix}$ or $\begin{vmatrix} 1 & x_1^2 \\ 1 & x_2^2 \end{vmatrix}$. The first is not zero because of $x_1 \neq x_2$, and the second equals, for a field of characteristic 2, $(x_1 - x_2)^2$.

We now describe a similar method, given in [11], for the construction of a $[\lambda s^2, \lambda s, s, 2]$, where λ and s are powers of the same prime, say $\lambda = p^u$, and $s = p^v$.

The elements of a $GF(p^{u+v})$ can be represented as $x_j = \sum_{i=0}^{u+v-1} a_i x^i$, with coefficients from $GF(p)$. To each x_j, let there correspond the sum of the first v terms, i.e. $y_j = \sum_{i=0}^{v-1} a_i x^i$, so that each y_j corresponds to $p^u = \lambda$ different x_j.

Write down the multiplication table of the x_j. Amongst the differences of corresponding elements in any two rows each element of $GF(p^{u+v})$ appears once, and if we replace x_j by y_j then each difference appears λ times.

Now replace each element of the modified multiplication table (i.e. after replacing x_j by y_j) by its row in the addition table of the y_j. The result is a $[\lambda s^2, \lambda s, s, 2]$. For the proof, which is quite simple, we refer the reader to [11]. A further row can be added which consists successively of λs times the first, λs times the second, ..., λs times the sth element of $GF(p^v)$.

Example

$p = 2$, $u = v = 1$, $\lambda = 2$, $s = 2$. We construct a $[8, 5, 2, 2]$.
Consider

$$x_j \quad 0 \quad 1 \quad x \quad 1+x$$

and hence

$$y_j \quad 0 \quad 1 \quad 0 \quad 1$$

The multiplication table of the x_j, i.e.

$$
\begin{array}{cccc}
0 & 0 & 0 & 0 \\
0 & 1 & x & 1+x \\
0 & x & 1+x & 1 \\
0 & 1+x & 1 & x
\end{array}
$$

is then "modified" into

$$
\begin{array}{cccc}
0 & 0 & 0 & 0 \\
0 & 1 & 0 & 1 \\
0 & 0 & 1 & 1 \\
0 & 1 & 1 & 1
\end{array}
$$

and the array becomes

```
01010101
01100110
01011010
01101001
```

to which can be added   00001111.

If $u \geqslant v$, then at least one more row can now be added. We refer to [11] and merely mention that altogether $c = 1+\lambda(s^{w+1}-1)/(s^w - s^{w-1})$ rows can be obtained, where w is the largest integer contained in u/v. In our example, the two further rows of the array are

```
00110011
00111100.
```

Maximal strength

It is known (see [129]) that m_t, that is the largest number of points in a $PG(m-1, s)$, such that no t of them lie in a flat of dimension smaller than $t-1$, satisfies the inequalities

$$
s^m \geqslant \sum_{i=0}^{h} \binom{m_t}{i} (s-1)^i \quad \text{when } t = 2h
$$

and

$$
s^m \geqslant \sum_{i=0}^{h} \binom{m_t}{i} (s-1)^i + \binom{m_t - 1}{h}(s-1)^{h+1} \quad \text{when } t = 2h+1.
$$

We shall now show that analogous inequalities hold for the largest number of rows in any orthogonal array, whether derived from a $EG(m,s)$ or not. We have to replace s^m by N.

Suppose N, s, and t to be given. We want to find the largest possible value of c, say $f(N, s, t)$.

Let a square matrix of order s be given, say (a_{ij}), $i, j = 0, ...,$ $s-1$, and consider the products $F(i_1, ..., i_c) = \left(\sum\limits_{j=0}^{s-1} a_{i_1 j} y_{1j} \right) \cdots$ $\left(\sum\limits_{j=0}^{s-1} a_{i_c j} y_{cj} \right)$. These are sums of s^c terms $a_{i_1 j_1} ... a_{i_c j_c} y_{1j_1} ... y_{cj_c}$, each term being determined by the ordered set $j_1, ..., j_c$, called an *assembly*.

Consider also an orthogonal array $[N, c, s, t]$ and retain in $F(i_1, ..., i_c)$ only those terms which correspond to assemblies given by the columns of the array. We call the resulting expression $G(i_1, ..., i_c)$. Take two of these, say $G(u_1, ..., u_c)$ and $G(v_1, ..., v_c)$ as functions of the products $y_{1j_1} ... y_{cj_c}$. Two of them are orthogonal if

$$ S = \Sigma (a_{u_1 j_1} ... a_{u_c j_c})(a_{v_1 j_1} ... a_{v_c j_c}) = 0, $$

where the summation extends over all assemblies from the given orthogonal array.

If the array is of strength c and hence contains all possible s^c assemblies λ times, then

$$ S = \lambda (a_{u_1 1} a_{v_1 1} + ... + a_{u_c c} a_{v_c c})^c . $$

If the matrix (a_{ij}) is orthogonal, then we see that $S = 0$, whichever $\{u_1, ..., u_c\}$ and $\{v_1, ..., v_c\}$ were chosen, unless $u_1 = v_1, ..., u_c = v_c$.

The case which interests us more is that where $t < c$. Therefore we assume that $a_{00} = ... = a_{0s-1} = 1$, and also that both the sets $\{u\}$ and $\{v\}$ contain 0 in $w = c - t$ common positions. Then the w^2 factors in each term of S, which refer to those common positions, are equal to 1, and we have

$$ S = \lambda (a_{u_1 1} a_{v_1 1} + ... + a_{u_c c} a_{v_c c})^t = 0. $$

This remains true if $w = c - t' > c - t$, because an array of strength t is also of strength t' ($< t$). We have thus shown

Theorem III.4 Any two expressions $G(u_1, ..., u_c)$ and $G(v_1, ..., v_c)$, where the sets $\{u\}$ and $\{v\}$ differ in not more than t arguments, while the other arguments are 0 in both expressions, are orthogonal.

We turn now to the question of how to find a set of functions $G(i_1, ..., i_c)$ which are, by virtue of Theorem III.4, pairwise orthogonal. We distinguish two cases.

Case (i): even t ($= 2h$).

We take all those $G(i_1, \ldots, i_c)$ where not more than h arguments differ from 0. Their number is

$$
\begin{array}{ll}
1 & \text{(all arguments equal to 0)} \\
+ \, c(s-1) & \text{(all but one equal to 0)} \\
+ \binom{c}{2}(s-1)^2 & \text{(all but two equal to 0)} \\
\multicolumn{2}{c}{- - - - - - - -} \\
+ \binom{c}{h}(s-1)^h. &
\end{array}
$$

Because the $G(i_1, \ldots, i_c)$ are linear combinations of terms $y_{1j_1} \ldots y_{cj_c}$, the total number of pairwise orthogonal $G(i_1, \ldots, i_c)$ cannot exceed the number of assemblies in the orthogonal array, so that

$$
N \geqslant \sum_{i=0}^{t/2} \binom{c}{i}(s-1)^i.
$$

Case (ii): odd $t \; (= 2h+1)$.

We take all those $G(i_1, \ldots, i_c)$ where not more than h arguments differ from 0, as in (i), and to this set we add those $G(i_1, \ldots, i_c)$ where i_1, say, differs from 0, and so do h others of the arguments. There will still be, in any two G, not more than t arguments in which they differ, while the other arguments are 0 in either. The number of the additional functions G is $\binom{c-1}{h}(s-1)^{h+1}$, and hence

$$
N \geqslant \sum_{i=0}^{h} \binom{c}{i}(s-1)^i + \binom{c-1}{h}(s-1)^{h+1}, \quad \text{where } h = \tfrac{1}{2}(t-1).
$$

Example

[4, 3, 2, 2] orthogonal matrix $\begin{pmatrix} 1 & 1 \\ 1 & -1 \end{pmatrix} = \begin{pmatrix} a_{00} & a_{01} \\ a_{10} & a_{11} \end{pmatrix}$

Coefficients of	$y_{00} \, y_{10} \, y_{20}$	$y_{00} \, y_{11} \, y_{21}$	$y_{01} \, y_{10} \, y_{21}$	$y_{01} \, y_{11} \, y_{20}$
$G(0 \; 0 \; 0)$	$a_{00} a_{00} a_{00}$	$a_{00} a_{01} a_{01}$	$a_{01} a_{00} a_{01}$	$a_{01} a_{01} a_{00}$
$G(0 \; 0 \; 1)$	$a_{00} a_{00} a_{10}$	$a_{00} a_{01} a_{11}$	$a_{01} a_{00} a_{11}$	$a_{01} a_{01} a_{10}$
$G(0 \; 1 \; 0)$	$a_{00} a_{10} a_{00}$	$a_{00} a_{11} a_{01}$	$a_{01} a_{10} a_{01}$	$a_{01} a_{11} a_{00}$
$G(1 \; 0 \; 0)$	$a_{10} a_{00} a_{00}$	$a_{10} a_{01} a_{01}$	$a_{11} a_{00} a_{01}$	$a_{11} a_{01} a_{00}$

Compare $G(0 \; 0 \; 0)$ and $G(0 \; 0 \; 1)$:

$$
a_{00} a_{10} + a_{01} a_{11} + a_{01} a_{11} + a_{00} a_{10} = 2(a_{00} a_{10} + a_{01} a_{11}) = 0.
$$

Compare $G(0\ 0\ 1)$ and $G(0\ 1\ 0)$:

$$(a_{00}a_{10})(a_{10}a_{00}) + (a_{01}a_{11})(a_{11}a_{01}) + (a_{00}a_{11})(a_{10}a_{01}) +$$
$$+ (a_{01}a_{10})(a_{11}a_{00}) = (a_{00}a_{10} + a_{01}a_{11})^2 = 0.$$

The inequality above gives $4 \geqslant 1 + 3(2-1)$ and we see that, in fact, equality holds in this case. However, this is not always so, and for special cases closer upper bounds have been established, as we shall see.

For $t = 2$ we obtain $f(N, s, 2) \leqslant (N-1)/(s-1)$ from the above formula, and this inequality was also proved, independently, in [86] and in [108]. There exist N and s for which the upper limit is reached. We have seen, for instance, that arrays $[s^m, (s^m-1)/(s-1), s, 2]$ can be constructed from a $PG(m, s)$. For a set of orthogonal Latin cubes of the first order of side s we have $N = s^3$, and since $f(s^3, s, 2) = s^2 + s + 1$, a complete set consists of $s^2 + s - 2$ cubes as stated above (page 33).

For $t = 3$, we obtain $f(\lambda s^3, s, 3) \leqslant (\lambda s^2 - 1)/(s-1) + 1$. For $s = 2$ this reduces to 4λ, and we have seen earlier how to construct an $[8\lambda, c+1, 2, 3]$ from a $[4\lambda, c, 2, 2]$. We find now that if the latter has the largest possible number of rows ($c = 4\lambda - 1$), then the former has also the largest possible number ($c+1 = 4\lambda$).

In [27] K.A. Bush has found expressions for $f(s^t, s, t)$ by building up an array step by step. We merely quote here his results. He has established the following upper bounds for $f(s^t, s, t)$:

$s \leqslant t$	$s \geqslant t$	
	s even	s odd, and $t \geqslant 3$
$t + 1$ (equality, if s is a prime power)	$s + t - 1$	$s + t - 2$

We have seen how to construct an array $[s^t, s+1, s, t]$ from polynomials in $GF(s)$ when $s \geqslant t$; in that case s must be the power of a prime. Hence then $f(s^2, s, 2) = s + 1$, and $f(s^3, s, 3) = s + 1$ if s is odd. We have also seen that, if s is even, and $t = 3$, then $f(s^t, s, t) \geqslant s + 2$. From the table above it follows that equality then holds in the last case.

Bose and Bush, in [11], have improved on $f(\lambda s^3, s, 3) \leqslant (\lambda s^2 - 1)/(s-1) + 1$ when $\lambda - 1$ is not divisible by $s - 1$. Call the remainder b. Then the right-hand side can be reduced by the largest integer contained in $t+1$, where t is the (always existing) positive root of $t^2 - t(2b - 2s + 1) + b(b - s + 1) = 0$.

If $\lambda - 1 = a(s - 1)$ and $(s - 1)^2(s - 2)$ is not divisible by $as + 2$, then $f(\lambda s^3, s, 3) \leqslant (\lambda s^2 - 1)/(s - 1) - 1$. This is also shown in [11].

E. Seiden has proved in [107] and in [108] that $f(81, 3, 3) = 10$. The construction of the corresponding array was given earlier in this chapter (pages 35-36).

The following formula follows from the structure of an orthogonal array $[N, c, s, t]$. Let n_i be the number of columns, not counting the first, in which there are i coincidences, i.e. i values which equal the value in the same row of the first column. Then

$$\sum_{i=0}^{c} \binom{i}{h} n_i = \binom{c}{h}(\lambda s^{t-h} - 1), \quad (h = 0, 1, ..., t).$$

Proof: Take any h $(\leqslant t)$ rows of the array. This can be done in $\binom{c}{h}$ ways. Within h given rows, $\lambda s^{t-h} - 1$ other columns will be identical with the first. Amongst these, any column with i coincidences will be counted $\binom{i}{h}$ times among the columns considered. The number of the latter is the right-hand side in the formula to be proved. (Note that $\binom{i}{h} = 0$ when $i < h$.)

For $h = 0$ we obtain $\sum_i n_i = N - 1$, which is evident from first principles

MacNeish's theorem

In [28] K. A. Bush has shown how to construct an orthogonal array from given ones of equal strength. Let an $[N_1, c_1, s_1, t]$ and an $[N_2, c_2, s_2, t]$ be given, and denote the smaller of c_1 and c_2 by c. Let the first array be (a_{ij}) and the second (b_{ij}). We take as elements of the new array the $s_1 s_2$ pairs (a_{ef}, b_{gh}) and construct the array:

$$(a_{11}, b_{11}) ... (a_{1N_1}, b_{11}) (a_{11}, b_{12}) ... (a_{1N_1}, b_{12}) ... (a_{1N_1}, b_{1N_2})$$

$$- \quad - \quad - \quad - \quad - \quad -$$

$$(a_{c1}, b_{c1}) ... (a_{cN_1}, b_{c1}) (a_{c1}, b_{c2}) ... (a_{cN_1}, b_{c2}) ... (a_{cN_1}, b_{cN_2}).$$

It is easily seen that this is an array $[N_1 N_2, c, s_1 s_2, t]$. Combining it with other arrays, we can obtain a $[\prod N_i, c, \prod s_i, t]$, where $c = \min_i c_i$.

The result was first mentioned in [7]. From it we obtain

Theorem III.5   Let $N = p_1^{a_1} \dots p_r^{a_r}$. Then an orthogonal array $[N^2, c, N, 2]$ exists, where c is the smallest of $p_i^{a_i} + 1$.

This is so, because arrays $\left[p_i^{2a_i}, p_i^{a_i} + 1, p_i^{a_i}, 2 \right]$, i.e. complete sets of orthogonal L.Sq's do exist whenever the side is a prime power. (Cf. also [65].)

Theorem III.5 was proved by MacNeish in [62]. He conjectured that he had, in fact, proved that $c - 2$, as defined above, was the largest possible number of pairwise orthogonal L.Sq's. His proof of the latter proposition was shown in [61] to be erroneous, though the proposition itself might still have been correct. (See, however, the next section.)

Orthogonal Latin squares

We shall now concentrate on orthogonal arrays of strength 2 and index 1, i.e. on sets of pairwise orthogonal L.Sq's. To begin with, we consider the consequences of Theorem III.5 and of the conjecture of MacNeish.

If N is odd, then Theorem III.5 guarantees the existence of at least two orthogonal L.Sq's (a Graeco-Latin square), because then the smallest prime factor cannot be smaller than 3. If N is even and contains the factor 4, then this is also true. But if N is of the form $2(2t+1) = 4t+2$, then no Graeco-Latin square would exist if $\min(p_i^{a_i}-1)$ were indeed the largest possible number in a set. Euler conjectured that no Graeco-Latin square of side $4t+2$ existed, and this is obviously true for $t = 0$, and was proved in [126] for $t = 1$ by a method of exhaustive enumeration. However, we shall see that it is incorrect for all other values of t.

We shall first describe the breakthrough achieved by Parker in [81].

Let an orthogonal array $[k^2, c, k, 2]$ be given. If a b.i.b.d., D say, with parameters v, b, k, r and with $\lambda = 1$ exists, then an orthogonal array $[v^2, c-1, v, 2]$ can be constructed as follows.

The $[k^2, c, k, 2]$ is equivalent to a set of $c-2$ orthogonal L.Sq's of side k. We can write the elements, and hence the array, in such a way that in the latter the first k columns and the first 2 rows are as shown in the table on the following page.

Omitting the first row, and the first k columns, we obtain an array, $A:I$, say, such that in any two rows all $k(k-1)$ ordered combinations without repetition of elements appear precisely once. (C.R. Rao denotes

$$0 \; 0 \; \ldots \; 0 \quad 1 \; 1 \; \ldots \; 1 \quad \ldots \quad (k-1) \; (k-1) \; \ldots \; (k-1)$$

$0 \; 1 \; \ldots \; (k-1)$	$0 \; 1 \; \ldots \; (k-1) \; \ldots \; 0 \quad\quad 1 \quad \ldots \; (k-1)$
$0 \; 1 \; \ldots \; (k-1)$	
$-$	$[k(k-1),\, c-1,\, k,\, 2] : I$
$-$	
$0 \; 1 \; \ldots \; (k-1)$	

such an array by $[k(k-1),\, c-1,\, k,\, 2] : I$; see [92], where a different type of modified array is also defined.) This array can be divided into $(k-1)$ subarrays such that in every row of each subarray every symbol appears precisely once.

We take one block of D, and replace in $A:I$ the element x (a number) by that in the xth position of that block of D. We do this for all blocks and place the resulting arrays next to each other. Thus we obtain a matrix of $bk(k-1)$ columns and $c-1$ rows. There appear in it altogether v symbols, and in any two rows any ordered pair $\left| \begin{smallmatrix} y \\ z \end{smallmatrix} \right|$ $(y \neq z)$ appears precisely once, because y and z appear once together in the same block, say in its u_1th and u_2th position, and thus they were inserted in the final design where $\left| \begin{smallmatrix} u_1 \\ u_2 \end{smallmatrix} \right|$ appeared in $A:I$ when that block was considered. Since $\lambda = 1$, and therefore $bk(k-1) = vr(k-1) = v(v-1)$, we have obtained a $[v(v-1),\, c-1,\, v,\, 2] : I$. We call it $B:I$.

By adding a matrix of $c-1$ rows, all of them listing the v varieties of the b.i.b.d., we obtain finally a $[v^2,\, c-1,\, v,\, 2]$.

We can add one more row to this array, obtaining a $[v^2,\, c,\, v,\, 2]$, if $B:I$ can be subdivided into $v-1$ subarrays of v columns each, such that every subarray contains all v varieties precisely once in each row. This is the case if the b.i.b.d. D was resolvable or symmetric, assuming that in the latter case it was written as a Latin rectangle (Youden square). In these cases we can construct, from the $[v(v-1),\, c-1,\, v,\, 2]$ $:I$, a $[v^2,\, c,\, v,\, 2]$ in the same way as the $[k^2,\, c,\, k,\, 2]$ given at the beginning of this section can be reconstructed from its truncated part, the $A:I$. We have proved

Theorem III.6 If a b.i.b.d. with parameters v, b, k, r, and with $\lambda = 1$ exists and if an orthogonal array $[k^2,\, c,\, k,\, 2]$, i.e. a set of $c-2$ orthogonal L.Sq's of side k also exists, then a $[v^2,\, c-1,\, v,\, 2]$, i.e. a set of

$c-3$ orthogonal L.Sq's of side v can be constructed.

If the b.i.b.d. is symmetric, or resolvable, then a $[v^2, c, v, 2]$, i.e. a set of $c-2$ orthogonal L.Sq's of side v (as many as there were of side k) can be constructed.

We are, in particular, interested in those cases where $c-2$ is larger than p^a-1, when p^a is the highest prime power factor contained in v.

Let s be a prime power, $v = 1+s+s^2$, $k = s+1$. Then the b.i.b.d. whose existence is postulated in Theorem III.6 exists as a $PG(2,s)$, and is symmetric. If $s+1$ is also a prime power, then a $[k^2, k+1, k, 2]$ exists and hence also a $[v^2, k+1, v, 2]$, i.e. $k-1$ orthogonal L.Sq's of side v.

Both s and $s+1$ are prime powers when s is a Mersenne prime $(2^p-1$, where p is a prime), or when $s+1$ is a Fermat prime $(2^{2^a}+1)$. If $s > 3$, then in either case $s \equiv 1$ (mod 3), and hence $1+s+s^2 \equiv 3$ (mod 9). MacNeish's conjecture leads to two orthogonal L.Sq's of side $1+s+s^2$, while Theorem III.6 gives s.

If $1+s+s^2$ is itself a prime power, then a set of as many as $s+s^2$ orthogonal L.Sq's exists, which is more than s. The smallest value which gives something new is $s = 4$, when $1+s+s^2 = 21$, not a prime power. A set of 4 orthogonal L.Sq's of side 21 is exhibited in [81]. It is the first counter-example to the conjecture of MacNeish.

We have seen (page 10) that a resolvable b.i.b.d. with $v = \binom{s}{2}$, $k = \frac{1}{2}s$, $\lambda = 1$ exists if $s = 2^m$. Let $m = 2n$, then $v = 2^{2n-1}(2^n+1)(2^n-1)$, $k = 2^{2n-1}$, so that $2^{2n-1}-1$ orthogonal L.Sq's of side v exist, while MacNeish's conjecture gives only 2^n-2 ([21]). This is true for all integer values of n.

We have also seen (page 11) that a resolvable b.i.b.d. with $v = s^3+1$, $k = s+1$, $\lambda = 1$ exists if s is a prime power. Hence there exist as many orthogonal L.Sq's with side s^3+1 as there are for side $s+1$.

Example (from [21])

$s = 31$, $k = 32 = 2^5$, $v = 29792 = 19 \times 32 \times 49$. There are at least 31 orthogonal L.Sq's of side v, which is more than 18 $(= 19-1)$.

No counter-example to Euler's conjecture has been produced in this way. To prove that this conjecture is false, further investigation is required. (The counter-example on p.31 was constructed only later.)

For a design which is balanced, i.e. in which each pair of varieties appears precisely once within the same block, and which contains blocks of sizes $k_1, ..., k_m$ out of v varieties altogether, we write $(v; k_1, ..., k_m)$. It has been generalised, but we shall not deal with the more general cases.

The falsity of Euler's conjecture

If there exist arrays $[k_i^2, c_i, k_i, 2]$ for each i, then an array $[v^2, c-1, v, 2]$ can be obtained, where c is the smallest of the c_i. For the purpose of its construction, we treat the $[k_i^2, c_i, k_i, 2]$ as we treated the arrays of k^2 columns before, put the resulting arrays next to each other, and retain only $c-1$ rows. It is easily seen that we thus obtain a $[v(v-1), c-1, v, 2] : I$ which we can expand into a $[v^2, c-1, v, 2]$.

It is shown in [23] that we can, in fact, obtain $c-1$ rows where c is the smallest of $c_1 + 1, ..., c_r + 1, c_{r+1}, ..., c_m$, if the blocks of sizes $k_1, ..., k_r$ are such that no two of them, of any of these sizes, have any variety in common. In this case we use, instead of a $[k_i(k_i - 1), c_i - 1, k_i, 2] : I$ $(i = 1, ..., r)$ the orthogonal array $[k_i^2, c_i, k_i, 2]$, and the matrix of $c-1$ rows which we added in the earlier case will now only contain those elements, if any, which do not already appear in the blocks of sizes $k_1, ..., k_r$.

Example

$(v; k_1, k_2) = (6, 3, 2)$. The blocks are given by the columns:

$$
\begin{array}{cccc} \quad & \quad & \quad & \quad \end{array}
$$

0	1	2	4		3	5	0
1	2	3	5		4	1	2
3	4	5	0				

Any two blocks of size 2 are such that they have no variety in common.

There exist arrays $[9, 4, 3, 2]$ and $[4, 3, 2, 2]$, and they give

012012   0011
120201   and 0101   respectively
201120   0110

From the latter arrays we obtain

013013	124124	235235	450450	3344	5551	0022
130301	241412	352523	504045	3434	5111	0202
301130	412241	523352	045504	3443	5155	0220

No further columns have to be added, because the blocks of size two already contain all six variables between them. (This example merely illustrates the procedure. A [36, 3, 6, 2] is equivalent to one single L.Sq. and, in fact, no pair orthogonal to it exists.)

Yet one more row can be added if the design from which we start is *separable* (a concept introduced in [21]), i.e. if any set of blocks of equal size can be divided into subsets such that each variety occurs in a subset the same number of times. If this number is unity, then we have a generalisation of a resolvable b.i.b.d.; if it is larger than one, then that of a symmetric b.i.b.d.

The proof that a further row can be added is analogous to that in the earlier case.

Balanced designs of the type we have been considering can be obtained by adding or by omitting varieties. First, take a case of addition.

Consider the resolvable b.i.b.d. with $v = 15$, $b = 35$, $r = 7$, $k = 3$, $\lambda = 1$:

1	2	3	2	4	6	4	3	7	3	6	5	5	1	4
4	12	8	3	9	13	6	14	10	7	11	12	2	10	11
7	9	14	5	14	11	1	11	8	2	8	13	6	12	9
6	11	13	7	8	10	5	13	12	1	10	9	3	14	8
5	10	15	1	12	15	2	9	15	4	14	15	7	13	15

6	7	1	7	5	2
5	8	9	1	13	14
4	13	10	3	10	12
2	12	14	4	9	11
3	11	15	6	8	15

Add to each block of the ith set of 5 blocks (each set contains all varieties) a new variety x_i, and add a further block $(x_1, x_2, ..., x_7)$. This produces a (22; 7, 4). There exist arrays [16, 5, 4, 2] and [49, 8, 7, 2]. Hence an orthogonal array [22^2, 4, 22, 2] can be found.

This is the first known counter-example to Euler's conjecture. The Graeco-Latin square is given in [20] and also in [21].

The existence of an infinite series of Graeco-Latin squares of sides $36m + 22$ is proved in [21], as follows.

It is known (see Chapter X of [98]) that resolvable b.i.b.d's exist with parameters $v = 6t + 3$, $b = (2t + 1)(3t + 1)$, $r = 3t + 1$, $k = 3$, $\lambda = 1$. If $t = 4m + 2$, then we have $v = 24m + 15$, $b = (8m + 5)(12m + 7)$, $r = 12m + 7$, $k = 3$, $\lambda = 1$.

To each block of the ith replication, i.e. the ith set of $8m + 5$ blocks, add a new variety, and combine all the new varieties in an additional block. This produces a $(36m + 22; 12m + 7, 4)$.

There exist arrays $[16, 5, 4, 2]$ and $[(12m+7)^2, 6, 12m+7, 2]$ because the smallest possible prime factor of $12m+7$ is 5. Hence at least two orthogonal L.Sq's of side $36m + 22$ exist.

Yet another counter-example to Euler's conjecture can be exhibited by starting from a resolvable b.i.b.d. with $v = s^m$, $b = s^{m-1}(s^m - 1)/(s - 1)$, $r = (s^m - 1)/(s - 1)$, $k = s$, $\lambda = 1$, where s is a prime power. These b.i.b.d's can be constructed from a $EG(m,s)$ by taking lines as blocks. Add another variable to the blocks of the first (and no other) replication, to obtain a $(s^m + 1; s + 1, s)$. If s as well as $s + 1$ are prime powers, then an $[(s^m + 1)^2, s, s^m + 1, 2]$ can be found.

If, in particular, s is $2^q - 1$ and a prime number larger than 3, and $m = 2n$, then $s^m + 1 = (2^q - 1)^{2n} + 1 \equiv 2 \pmod 4$, so that once more at least two orthogonal L.Sq's exist of a side of the form $4t + 2$.

We now give an example of the construction of a balanced design by omitting varieties. We start from the b.i.b.d. which is equivalent to a $PG(2, s)$. Let $s = 16$, and omit six treatments from all the blocks where they appear. This gives a $(s^2 + s + 1 - 6; s + 1, s, s - 5)$, i.e. a $(267; 17, 16, 11)$. Therefore a set of at least 9 orthogonal L.Sq's of side 267 exists, while MacNeish's conjecture gives only two, since $267 = 3 \times 89$.

We mention one more scheme, and because we shall make use of it later, we express it as a theorem.

Theorem III.7 Denote the largest number of L.Sq's of size n in an orthogonal set by $N(n)$. Then $N((c - 1)m + x)$ is at least as large as the smallest of

$$N(c) - 1, \quad N(c - 1) - 1, \quad N(x), \quad \text{and} \quad N(m),$$

provided $x < m$.

Proof: Let an $[m^2, c, m, 2]$ be given and omit the first row. The remaining array can be divided into m subarrays such that in every row of each subarray each symbol (let them be $0, 1, ..., m - 1$) occurs precisely once. Replace the symbol i in the jth row by a new variety $x_{(j-1)m+i}$. Take the columns as blocks of a design. This is a configuration with $v = (c - 1)m$, $b = m^2$, $k = c - 1$, $r = m$, but not balanced. (In fact, it is a partially balanced incomplete block design of group-divisible type; for these, see Chapter V.) To make it balanced, we add m further blocks, each consisting of

$$x_{(j-1)m+0}, \ldots, x_{(j-1)m+m-1}, \quad \text{for} \quad j = 1, \ldots, m.$$

We illustrate this procedure with $m = 3$, $c = 4$. After omitting the first row of a $[9, 4, 3, 2]$ we obtain

012	012	012
012	120	201
012	201	120

and from it the design

036	048	057
147	156	138
258	237	246

To this we add the three blocks

$$012 \qquad 345 \qquad 678.$$

To obtain the balanced design we want, we add a new variety to each of the first x $(<m)$ replications and combine all these new varieties in a further block. The result is a $((c-1)m+x; c, c-1, x, m)$. Notice that no two of the added $m+1$ blocks contain common varieties. From this the Theorem follows.

Example

$$m = 11, \quad c = 9, \quad x = 7.$$

$$N(95) \geqslant \min(N(9)-1, N(8)-1, N(7), N(11)) = \min(7, 6, 6, 10) = 6.$$

In [83] E.T. Parker gave the following example of a Graeco-Latin square of side 10:

0417298365	0786935412
8152739406	6178094523
9826374510	5027819634
5983047621	9613782045
7698415032	3902478156
6709852143	8491357260
3071986254	7859246301
1234560789	4560123789
2345601897	1234560978
4560123978	2345601897

The last three rows and columns contain two orthogonal L.Sq's of side 3, and because $10 > 3^2$, but not $\geqslant 3^2+3$, no third L.Sq. exists which is orthogonal to the two squares given.

The same Graeco-Latin square is reproduced, with the rows in a different order, in [23]. The latter paper contains a somewhat simpler argument for its construction than Parker's original demonstration, and we reproduce it here.

To begin with, form a $[(3m+1)^2, 4, 3m+1, 2]$ for odd m, as follows. We define the matrix A_0 as

0	0	...	0	1	2	...	m	$2m$	$2m-1$	...	$m+1$	x_1	x_2	...	x_m	
1	2	...	m	0	0	...	0	x_1	x_2	...	x_m	$2m$	$2m-1$	...	$m+1$	
$2m$	$2m-1$	...	$m+1$	x_1	x_2	...	x_m	0	0	...	0	1	2	...	m	
x_1	x_2	...	x_m	$2m$	$2m-1$	...	$m+1$	1	2	...	m	0	0	...	0	

where the x_i are indeterminate symbols.

Form A_j by adding j $(0, 1, ..., 2m)$ to the elements of A_0, except to the x_i which remain unaltered. Reduce mod $2m+1$. In the matrix $(A_0, A_1, ..., A_{2m})$ any ordered pair of two distinct residuals of $2m+1$, or any ordered pair consisting of one of the residuals and one of the x_i, occurs once in any two-rowed submatrix. To obtain an orthogonal array, we add the matrix consisting of four identical rows 0 1 ... $2m$, and also an orthogonal array $[m^2, 4, m, 2]$, which exists because m is odd. In the latter array we use the x_i as elements.

The number of all the columns we shall then have is $4m(2m+1) + (2m+1) + m^2 = (3m+1)^2$, and they form an orthogonal array. When $m = 4n+3$, then this array can be written $[(12n+10)^2, 4, 12n+10, 2]$.

We have thus obtained an infinite series of Graeco-Latin squares of sides $4t+2$, and 10 is one of these numbers.

The complete theorem on Graeco-Latin squares

We turn now, finally, to the proof that Graeco-Latin squares exist for all sides except 2 and 6. The proof starts with listing various cases, and we give only the gist of the argument.

[20] and [23] contain lists of lower bounds (not necessarily the best) of possible numbers of orthogonal L.Sq's in a set which have been proved for various values of n, using methods outlined above. These numbers are, in all cases, at least 2, and cover all sides larger than 6, up to 726 inclusive.

All numbers of the form $4t+2$ which are larger than 726 can be written $4a+144b+10$, where $0 \leqslant a \leqslant 35$ and $b \geqslant 5$. In Theorem III.7, take $c = 5$, $m = 36b$, and $x = 4a+10$. Then $m \geqslant 180$ and $x \leqslant 150$, so that $x < m$. It follows that

$$N(4 \times 36b + 4a + 10) \geqslant \min[N(5) - 1, N(4) - 1, N(4a + 10), N(36b)].$$

Now $N(5) - 1 = 3$, $N(4) - 1 = 2$, $N(4a + 10) \geqslant 2$, and $N(36b) \geqslant 3$, because the smallest prime factor of $36b$ is 4, in view of $b \geqslant 5$. Hence $N(4a + 144b + 10) \geqslant 2$, which proves

Theorem III.8: Graeco-Latin squares exist for all sides except for side 2 and for side 6.

In [84] Parker mentions that he has found many Graeco-Latin squares of side 10 by trials on a computer, and he shows one "totally unlike the examples obtained in 1959 by purely mathematical methods". But he did not find any set of three orthogonal L.Sq's of that side.

On the other hand, N.S. Mendelsohn and others have found four sets of five orthogonal L.Sq's of side 12, but no larger set of this side (see [67]).

In [32] the authors have proved by methods of number theory that the maximal number of pairwise orthogonal L.Sq's of side n tends, with increasing n, to infinity. There exists a number n_0 such that $N(n)$ exceeds $\dfrac{n^{1/91}}{3}$ for all $n > n_0$, but they remark that $1/91$ is far from being the best exponent.

To end this chapter, we mention as a matter of interest that Rojas and White discuss in [97] designs of $a.b$ varieties in $a.b$ columns and **b** rows, such that each variety occurs in every row, and once in every group of a columns. They call these designs "modified L.Sq's", and they are the same as Yates's "semi-Latin squares" (see [130]). We give here an example:

1	2	3	4	5	6	7	8	9
4	5	6	7	8	9	1	2	3
7	8	9	1	2	3	4	5	6.

EXERCISES

(1) Construct two orthogonal Latin squares using the following points and lines of a $PG(2,3)$:

Points:

A (100)	B (010)	C (001)	a (211)	b (121)	c (112)
Q (120)	R (102)	S (012)	X (011)	Y (101)	Z (110)
P (111)					

Lines:

$x_0 + x_2 \quad = 0,$	$x_0 + 2x_2 \quad = 0,$	$x_2 \qquad = 0$	(through B)
$acRB$	$bPYB$	$QZAB$	
$x_0 + x_1 \quad = 0,$	$x_0 + 2x_1 \quad = 0,$	$x_1 \qquad = 0$	(through C)
$abQC$	$cPZC$	$RYAC$	
$x_1 + x_2 \quad = 0,$	$x_1 + 2x_2 \quad = 0,$	$x_0 + x_1 + x_2 = 0$	(through S)
$AbcS$ (1)	$aYZS$ (0)	$PQRS$ (2)	
$x_0 + x_1 + 2x_2 = 0,$	$x_0 + 2x_1 + x_2 = 0,$	$x_1 + 2x_2 \quad = 0$	(through X)
$QYcX$ (2)	$RZbX$ (1)	$APaX$ (0)	

(2) Construct a complete set of orthogonal Latin squares of side 4.

(3) Construct a complete set of orthogonal Latin squares of side 5.

(4) Extend the following rectangles into Latin squares of side 5:

		012		412
(i)		340	(ii)	341
		123		123

CHAPTER IV

PARTIALLY BALANCED INCOMPLETE BLOCK DESIGNS

Definition

In this chapter we deal with a generalisation of b.i.b.d's. Let v varieties be given. We define an association scheme as follows: any two varieties are either first, or second..., or mth associates, and this relationship is symmetric. Each variety has n_i different ith associates. Given any two varieties that are ith associates, there are $p_{jk}^i = p_{kj}^i$ varieties which are jth associates of the one and kth associates of the other; these values do not depend on which pair of ith associates we started from. (The cases $i = j$ or $i = k$ are not excluded from this definition.)

We simplify our formulae by taking each variety to be the 0-th associate of itself, and only of itself. Hence $n_0 = 1$, $p_{jk}^0 = n_j$ when $j = k$, and $= 0$ otherwise, while $p_{0k}^i = 1$ when $i = k$, and $= 0$ otherwise.

We define a partially balanced incomplete block design (p.b.i.b.d.) to be a b.i.b.d., except that not all pairs of varieties appear equally often within the same block. Instead, any two varieties which are ith associates appear together in λ_i blocks. The λ_i need not all be different, but if they are all equal (or if $m = 1$), then the design reduces to a b.i.b.d. We also use λ_0 as an alternative to r, the number of repetitions of any variety.

These designs were introduced in [16]. For a broadening and rephrasing of the original definitions see [73] and [19].

The p_{jk}^i and n_i define the association scheme, but they do not define the design by themselves. The parameters v, b, r, k, λ_i and n_i are called of the first kind, and the p_{jk}^i parameters of the second kind. There are altogether $2m + 4$ of the former (not counting n_0), and $\frac{1}{2}(m + 1)^2 (m + 2)$ of the latter.

We have $n_0 + n_1 + \ldots + n_m = v$ and $n_0 \lambda_0 + \ldots + n_m \lambda_m = rk$, because whenever a variety appears (which is r times), it appears together with $k - 1$ others in the same block, λ_i times with each of its ith associates, and $n_0 \lambda_0 = r$.

Further relations between the parameters can be derived as follows:—

Take any two varieties, say v_1 and v_2, which are ith associates. The jth $(i \neq j)$ associates of v_1 number n_j, and any of these are either a first, or second, $\ldots$, or mth associate of v_2. Hence $p_{1j}^i + \ldots + p_{mj}^i = n_j$.

This formula holds also for $i = 0$, when it reduces to $p_{jj}^0 = n_j$, or for $j = 0$, when it means $p_{0i}^i = n_0 = 1$.

We shall now prove that $n_i p_{jk}^i = n_j p_{ik}^j = n_k p_{ij}^k$. Consider a variety, say v_1. It has n_i different ith associates, say $x_1, x_2, \ldots, x_{n_i}$ and n_j different jth associates, say $y_1, y_2, \ldots, y_{n_j}$.

Every one of the x's is related to p_{jk}^i of the y's which are its kth associates. We thus obtain $n_i p_{jk}^i$ "connections". But the relationship of being kth associates is symmetric. Therefore, the same connections are obtained if we start from the y's and relate each of these to the x's which are its kth associates. There are then $n_j p_{ik}^j$ connections, and their equality establishes the formulae to be proved.

These relationships establish constraints on the choice of the parameter values. Altogether, given the n_i, only $m(m^2 - 1)/6$ of the p_{jk}^i are independent. (In this count we exclude 0-th associates, since their parameters are all fixed in advance.)

Proof: To begin with, consider the matrix (p_{jk}^i) for some fixed i, say for $i = 1$. Since the matrix is symmetric, it is defined by $\binom{m+1}{2}$ values; since the row and column totals are fixed, only $\binom{m}{2}$ of these are independent.

If the p_{2k}^1 are known, then the relation $n_1 p_{2k}^1 = n_2 p_{1k}^2$ determines all $p_{1k}^2 = p_{k1}^2$. Hence in the matrix (p_{jk}^2) only the second to mth columns and rows remain free to be determined, and these form a matrix in which $\binom{m-1}{2}$ values are independent, since row and column totals are again fixed.

Continuing, we find $\binom{m-2}{2}$ independent values among the p_{jk}^3, and so on, until $\binom{m-(m-2)}{2} = 1$ value remains among the p_{jk}^{m-1}. The values of the p_{jk}^m are then implicitly fixed and we have in all $\sum_{i=0}^{m-2} \binom{m-i}{2} = m(m^2 - 1)/6$ independent values of the parameters of the second kind.

(The last formula is easily verified by induction.)

This completes the proof.

For $m = 2$, we have $m(m^2 - 1)/6 = 1$, and this will be used in the next chapter.

Even when the parameters satisfy the necessary conditions which we have established, it is not certain that a p.b.i.b.d. with such parameters exists. For instance, let $v = 13$, $b = 13$, $r = 4$, $k = 4$, $\lambda_1 = 2$, $\lambda_2 = 0$, $n_1 = 6$, $n_2 = 6$.

$$(p_{jk}^1) = \begin{pmatrix} 2 & 3 \\ 3 & 3 \end{pmatrix}, \quad (p_{jk}^2) = \begin{pmatrix} 3 & 3 \\ 3 & 2 \end{pmatrix}.$$

For such a design to exist, there must be two blocks to contain two given varieties which are first associates ($\lambda_1 = 2$), say x and y. The other two elements in either block must be first associates of both x and y ($\lambda_2 = 0$). But only two such elements exist, and therefore the two blocks would be identical. Even if we allowed this to happen, it would be impossible in this case, because b is odd, and cannot equal the number of blocks if each appears twice. (This example is taken from [35].)

Construction

We can construct p.b.i.b.d's by selecting blocks from suitable b.i.b.d's. In Chapter II we considered b.i.b.d's whose blocks were hyperplanes in parallel pencils in a $EG(m,s)$. This time we take only the m pencils $x_i = c_i$ ($i = 0, 1, \ldots , m-1$), where the c_i are the elements of $GF(s)$. The points are varieties, and we define as ith associates any two points with $m - 1$ equal co-ordinates.

We have $v = s^m$, $b = m.s$, $k = s^{m-1}$, $r = m$. The ith associates of the point $(x_1, \ldots , x_m) = x$, say, are found by selecting $m - i$ co-ordinates — this can be done in $\binom{m}{i}$ ways — and giving to each of the remaining co-ordinates any value different from that in x — which can be done in $(s-1)^i$ different ways. Therefore $n_i = \binom{m}{i}(s-1)^i$. Two varieties which are ith associates and have, therefore, $m - i$ equal co-ordinates, appear together in all blocks defined by any one of those co-ordinates. Hence $\lambda_i = m - i$.

Now to the parameters of the second kind. If two varieties, say x and y, are ith associates, then they have $m - i$ co-ordinates with equal values, and i co-ordinates with unequal ones. A variety which is a jth associate of x and a kth associate of y will have $m - j$ co-ordinates in common with x and $m - k$ in common with y. Of the $m - j$ former ones, there will be u (any value from 0 to $m - i$) taken from those which are common to x and y, and $m - j - u$ from the remaining i co-ordinates of x, while $m - k - u$ will be among the

$m - (m - i) - (m - j - i) = i + j + u - m$ co-ordinates of y which do not have common values in x and y, and which have not been chosen from among those that are not common. The number of all these varieties is $\binom{m-i}{u} \binom{i}{m-j-u} \binom{i+j+u-m}{m-k-u}$.

Each of the remaining $m-i-u$ co-ordinates out of those common to x and y can have any of $s-1$ values, and each of the other remaining co-ordinates, numbering $i-(m-j-u)-(m-k-u) = i+j+k-2m-2u = a$, say, can take any of $s-2$ values (not those of either x or y). Thus we obtain

$$p^i_{jk} = \sum_{u=0}^{m-1} \binom{m-i}{u} \binom{i}{m-j-u} \binom{i-m+j+u}{m-k-u} (s-1)^{m-i-u}(s-2)^a.$$

When $s = 2$, then only that term in this sum will differ from 0 in which $a = 0$, i.e. where $u = \frac{1}{2}(i+j+k-2m)$.

When $m = 2$, then $\lambda_1 = 1$, $\lambda_2 = 0$, $n_1 = 2(s-1)$, $n_2 = (s-1)^2$,

$$(p^1_{jk}) = \begin{pmatrix} s-2 & s-1 \\ s-1 & (s-1)(s-2) \end{pmatrix}, \quad (p^2_{jk}) = \begin{pmatrix} 2 & 2(s-2) \\ 2(s-2) & (s-2)^2 \end{pmatrix}.$$

This is a Latin square type scheme L_2 (see Chapter V).

When $m = 3$, $s = 2$, then we can imagine the design to be derived from a cube with vertices $1, 2, \ldots, 8$. Let each block contain vertices of the same face. Two vertices are first associates if they lie on the same edge, second associates if they are diagonally opposite on a face, and third if they lie at the two ends of a diagonal of the cube. The design is then 1234 2367 5678 1256 1458 3478, where $v = 8$, $b = 6$, $k = 4$, $r = 3$, $n_1 = n_2 = 3$, $n_3 = 1$, $\lambda_1 = 2$, $\lambda_2 = 1$, $\lambda_3 = 0$. The parameters of the second kind are

$$(p^1_{jk}) = \begin{pmatrix} 0 & 2 & 0 \\ 2 & 0 & 1 \\ 0 & 1 & 0 \end{pmatrix}, \quad (p^2_{jk}) = \begin{pmatrix} 2 & 0 & 1 \\ 0 & 2 & 0 \\ 1 & 0 & 0 \end{pmatrix}, \quad (p^3_{jk}) = \begin{pmatrix} 0 & 3 & 0 \\ 3 & 0 & 0 \\ 0 & 0 & 0 \end{pmatrix}.$$

These designs are special cases of "quasi-factorial" designs with $s_1 \times s_2 \times \ldots \times s_n$ varieties $(x_1, \ldots, x_n)$, $x_i = 1, \ldots, s_i$, and such that $(x_1, \ldots, x_n)$ and $(y_1, \ldots, y_n)$ appear together in $\lambda_{c_1 \ldots c_n}$ blocks with $c_i = 1$ if $x_i = y_i$, and $c_i = 0$ otherwise. The above designs are obtained by suitably collapsing the association scheme.

Other constructions of p.b.i.b.d's from finite geometries will be given in the next chapter.

[123] contains examples of the construction of p.b.i.b.d's from difference sets, when v is the power of a prime. Bose and Nair have already considered such constructions in [16] for $m=2$ and arbitrary values of v. The following example illustrates the procedure.

Take 1, 3, 9 (mod 13). The differences 2, 5, 6, 7, 8, 11 appear once each, and 1, 3, 4, 9, 10, 12 do not appear at all. The blocks $(1+t, 3+t, 9+t)$, $t=0,\ldots,12$ form a p.b.i.b.d. with $n_1 = n_2 = 6$, $\lambda_1 = 1$, $\lambda_2 = 0$,
$$(p_{jk}^1) = \begin{pmatrix} 2 & 3 \\ 3 & 3 \end{pmatrix}, \ (p_{jk}^2) = \begin{pmatrix} 3 & 3 \\ 3 & 2 \end{pmatrix}.$$

Another example starts with 1, 2, 3 (mod 5). We obtain a p.b.i.b.d. whose association scheme can be illustrated by a pentagram, where any two vertices connected by a line are second associates, and any other pair are first associates.

In [91] C.R. Rao mentions the following "circular lattice designs": take n concentric circles and draw n diameters. Let the intersections represent $2n^2$ varieties, and let the circles and diameters represent $2n$ blocks, with $k = 2n$ varieties in each. Every variety will be twice repeated. The association scheme is as follows: two points on diametrically opposed ends of the same diameter are first associates, two on the same circle, or on the same diameter, but not on both, are second associates, and all other pairs are third associates. Then $n_1 = 1$, $n_2 = 4(n-1)$, $n_3 = 2(n-1)^2$; $\lambda_1 = 2$, $\lambda_2 = 1$, $\lambda_3 = 0$;

$$(p_{jk}^1) = \begin{pmatrix} 0 & 0 & 0 \\ 0 & 4(n-1) & 0 \\ 0 & 0 & 2(n-1)^2 \end{pmatrix}, \quad (p_{jk}^2) = \begin{pmatrix} 0 & 1 & 0 \\ 1 & 2(n-2) & 2(n-1) \\ 0 & 2(n-1) & 2(n-1)(n-2) \end{pmatrix}$$

$$(p_{jk}^3) = \begin{pmatrix} 0 & 0 & 1 \\ 0 & 4 & 4(n-2) \\ 1 & 4(n-2) & 2(n-2)^2 \end{pmatrix}.$$

These designs are resolvable into two replications, one containing the n circles as blocks, and the other the n diameters.

In [93], P.V. Rao refers to [111] where three associate classes are treated and where it is shown that certain "partially balanced" matrices can be used to find p.b.i.b.d's with $2m+1$ associate classes.

Various association schemes

B. Harshbarger has developed designs (in [50] and [51]) some of which are p.b.i.b.d's. They were systematically studied by K.R. Nair in [72].

The "simple rectangular lattice" of $v = n(n-1)$ varieties may be described by allocating to each variety an ordered pair (x,y), $x \neq y$, $x,y = 1, \ldots, n$. We then let the first associates of (x,y) be the varieties which have either x or y in common with it (hence $n_1 = 2(n-2)$), second associates those which have no mark in common (hence $n_2 = (n-2)(n-3)$), third those with one mark in common, but not in the same position ($n_3 = 2(n-2)$), and fourth the variety (y,x) (hence $n_4 = 1$). The blocks are $(x,1)\ldots(x,n)$ for all x, and $(1,y)\ldots(n,y)$ for all y. Then $b = 2n$, $k = n-1$, $r = 2$, $\lambda_1 = 1$, $\lambda_2 = \lambda_3 = \lambda_4 = 0$;

$$(p_{jk}^1) = \begin{pmatrix} n-3 & n-3 & 1 & 0 \\ n-3 & (n-3)(n-4) & n-3 & 0 \\ 1 & n-3 & n-3 & 1 \\ 0 & 0 & 1 & 0 \end{pmatrix}$$

$$(p_{jk}^2) = \begin{pmatrix} 2 & 2(n-4) & 2 & 0 \\ 2(n-4) & (n-4)(n-5) & 2(n-4) & 1 \\ 2 & 2(n-4) & 2 & 0 \\ 0 & 1 & 0 & 0 \end{pmatrix}$$

$$(p_{jk}^3) = \begin{pmatrix} 1 & n-3 & n-3 & 1 \\ n-3 & (n-3)(n-4) & n-3 & 0 \\ n-3 & n-3 & 1 & 0 \\ 1 & 0 & 0 & 0 \end{pmatrix}$$

$$(p_{jk}^4) = \begin{pmatrix} 0 & 0 & 2(n-2) & 0 \\ 0 & (n-2)(n-3) & 0 & 0 \\ 2(n-2) & 0 & 0 & 0 \\ 0 & 0 & 0 & 0 \end{pmatrix}$$

These designs are also mentioned in [70].

When $n = 3$, then no second associates exist. Renumbering the third and fourth association to be second and third, we obtain a scheme with three classes, and the parameters of the second kind are then

$$(p^1_{jk}) = \begin{pmatrix} 0 & 1 & 0 \\ 1 & 0 & 1 \\ 0 & 1 & 0 \end{pmatrix}, \quad (p^2_{jk}) = \begin{pmatrix} 1 & 0 & 1 \\ 0 & 1 & 0 \\ 1 & 0 & 0 \end{pmatrix}, \quad (p^3_{jk}) = \begin{pmatrix} 0 & 2 & 0 \\ 2 & 0 & 0 \\ 0 & 0 & 0 \end{pmatrix}.$$

A "triple rectangular lattice" is constructed as follows:

Take an $[n(n-1), 3, n, 2] : I$ (defined in Chapter III) and number its columns from 1 to $n(n-1)$. The $3n$ blocks of the design are formed by all those numbers of columns which contain the same entry in the first, or in the second, or in the third row of the orthogonal array. Then $k = n-1$, and $r = 3$.

Nair has shown that there are only two p.b.i.b.d's among these lattices, namely those for $n = 3$ or $n = 4$.

$n = 3$	$n = 4$
$v = 6,\ b = 9,\ k = 2,\ r = 3$	$v = 12,\ b = 12,\ k = 3,\ r = 3$
$\lambda_1 = 1,\ \lambda_2 = 0$	$\lambda_1 = 1,\ \lambda_2 = \lambda_3 = 0$
$n_1 = 3,\ n_2 = 2$	$n_1 = 6,\ n_2 = 3,\ n_3 = 2$
$(p^i_{jk}) = \begin{pmatrix} 0 & 2 \\ 2 & 0 \end{pmatrix}, \begin{pmatrix} 3 & 0 \\ 0 & 1 \end{pmatrix}$	$(p^i_{jk}) = \begin{pmatrix} 2 & 2 & 1 \\ 2 & 0 & 1 \\ 1 & 1 & 0 \end{pmatrix}, \begin{pmatrix} 4 & 0 & 2 \\ 0 & 2 & 0 \\ 2 & 0 & 0 \end{pmatrix}, \begin{pmatrix} 3 & 3 & 0 \\ 3 & 0 & 0 \\ 0 & 0 & 1 \end{pmatrix}$

The association schemes can easily be recovered from the parameters.

Other p.b.i.b.d's with $r = 3$ are studied in [71].

Resolvable* p.b.i.b.d's are introduced in [18] as follows.

Consider the incidence matrix of a symmetric b.i.b.d. with v varieties and blocks, and k varieties in each block. Replace every cell with 1 by p different varieties, which are different in each cell, and every cell with 0 by q different varieties, again different in each cell. We have then $v(kp + vq - kq)$ varieties, $kp + vq - kq$ in each row of the matrix, and in each column. We take the rows, and the columns, as blocks and thus obtain $2v$ blocks, and each variety is twice repeated.

A *rectangular association scheme* is defined for $v = s.t$ varieties as a rectangle with s rows and t columns, with first associates in the same row, second in the same column, and all other pairs being third associates. Hence $n_1 = t-1$, $n_2 = s-1$, $n_3 = (s-1)(t-1)$, and

---

* Defined as for b.i.b.d's.

$$(p_{jk}^1) = \begin{pmatrix} t-2 & 0 & 0 \\ 0 & 0 & s-1 \\ 0 & s-1 & (s-1)(t-2) \end{pmatrix}, \quad (p_{jk}^2) = \begin{pmatrix} 0 & 0 & t-1 \\ 0 & s-2 & 0 \\ t-1 & 0 & (s-2)(t-1) \end{pmatrix}$$

$$(p_{jk}^3) = \begin{pmatrix} 0 & 1 & t-2 \\ 1 & 0 & s-2 \\ t-2 & s-2 & (s-2)(t-2) \end{pmatrix}.$$

Conversely, these parameters define the association scheme uniquely. Indeed, let x_1 and x_2 be first associates. Because $n_1 = t-1$ and $p_{11}^1 = t-2 = n_1 - 1$, the $s.t$ values can be grouped into s rows each of t elements, so that those in the same row, and only those, are first associates. This property is symmetric, and in the present case also transitive.

x_1 has $s-1$ second associates, and because $p_{12}^2 = 0$ there will be precisely one second associate in each row of which x_1 is not a member. We can therefore write pairs of second associates into the same column, since $p_{22}^2 = s-2 = n_2 - 1$.

A design constructed by taking as blocks the union of a row and a column of the rectangular association scheme has parameters $v = b = s.t$, $r = k = s+t-1$, $\lambda_1 = t$, $\lambda_2 = s$, $\lambda_3 = 2$. (This is a quasi-factorial design with $n = 2$.)

A different design results when the variety common to the row and the column united in a block is omitted.

In certain cases we can make an association scheme collapse into one with less classes. For instance, if in a rectangular association scheme $s = t$, and hence also $\lambda_1 = \lambda_2$, and if we then let first associates be any two varieties either in the same row or in the same column, while any other pair are second associates, we obtain the L_2 association scheme (see next chapter).

On the other hand, if we let two varieties in the same column be first, and any other two second associates, then we obtain a so-called "group-divisible" association scheme, which we study in detail in the next chapter.

We obtain yet another association scheme if we denote $v = r_1 \times \dots \times r_m$ varieties by $v_{i_1 \dots i_m}$, where $i_j = 1, 2, \dots r_j$. Let first associates be two varieties with all but the last subscripts equal, and generally $(m-i)$th associates those with the first i subscripts equal,

but the $(i+1)$th different. Then $n_1 = r_m - 1$, $n_2 = r_m(r_{m-1} - 1)$, ...,
$n_m = r_m r_{m-1} \cdots r_2(r_1 - 1)$.

This scheme was introduced in [100] and discussed in [102]. In [88] it is proved that the values of the parameters of the second kind define the association scheme uniquely. These designs are special cases of the intra- and inter-group b.i.b.d's defined in [129]. We do not discuss this scheme any further here. In the next chapter we shall deal with a special case, namely that of group-divisible designs.

Association matrices

In [15] Bose and Mesner use another method to describe an association scheme. They define it by "association matrices" B_i, of order v and symmetric:

$$B_i = \begin{pmatrix} b_{1i}^1 & b_{1i}^2 & \cdots & b_{1i}^v \\ - & - & - & - \\ b_{vi}^1 & b_{vi}^2 & \cdots & b_{vi}^v \end{pmatrix}, \quad i = 0, 1, \ldots, m,$$

where $b_{ti}^s = 1$ if the varieties s and t are ith associates, and $= 0$ otherwise. The total of the entries of each column, and also of each row of B_i, is n_i. Clearly,

(4.1) $B_0 = I_v$ (i.e. the identity matrix of order v). Furthermore, b_{ti}^s is unity for one and only one value of i when s and t are given, so that

(4.2) $\sum\limits_{i=0}^{m} B_i = J_v$ (i.e. the matrix of order v whose elements are all 1). A linear combination of the B_i can only be 0 if all the coefficients are 0.

It follows from the definition of the b_{ti}^s that

(4.3) $p_{jk}^i = \sum\limits_{t=0}^{m} b_{rt}^c \, p_{jk}^t$ if c and r are ith associates, because only one term in the sum of the right-hand side is different from 0.

Consider the matrix product $B_j B_k$. It contains in its rth row and cth column the sum $\sum\limits_{t=1}^{v} b_{rj}^t \, b_{tk}^c$. A term in this sum is 1 if and only if t and r are jth, and t and c are kth associates. Otherwise it is 0. Therefore, if r and c are ith associates, then

(4.4) $\sum\limits_{t=1}^{v} b_{rj}^t \, b_{tk}^c = p_{jk}^i$.

Comparing this result with (4.3), we have

(4.5) $\quad B_j B_k = \sum_{t=0}^{m} B_t \, p_{jk}^t = \sum_{t=0}^{m} B_t \, p_{kj}^t = B_k B_j$.

This formula has appeared in [127]. We have not assumed that j and k must be different, and therefore any power of B_i can also be expressed as a linear combination of the B_i. We have

$$B_i (B_j B_k) = B_i \sum_u p_{jk}^u B_u = \sum_u p_{jk}^u B_i B_u = \sum_u \sum_t p_{jk}^u p_{iu}^t B_t \,,$$

and also

$$(B_i B_j) B_k = \sum_u p_{ij}^u B_u B_k = \sum_u \sum_t p_{ij}^u p_{uk}^t B_t \,.$$

Matrix multiplication is associative, and the B_i are linearly independent, as we have pointed out. It follows that

(4.6) $\quad \sum_{u=0}^{m} p_{jk}^u p_{iu}^t = \sum_{u=0}^{m} p_{ij}^u p_{uk}^t \quad$ for $0 \leqslant i,j,k,t \leqslant m$.

For $t=0$, in particular, this means that $n_k p_{ij}^k = n_i \, p_{jk}^i$. Given a p.b.i.b.d. with incidence matrix A, we have

(4.7) $\quad AA' = \sum_{i=0}^{m} \lambda_i B_i$.

Conversely, let symmetric matrices of order v with elements 0 and 1 be given, satisfying (4.1), (4.2) and (4.5) with some constants p_{jk}^t. Then these constants are parameters of an association scheme.

We have to show that

(a) $p_{jk}^t = p_{kj}^t$;

(b) $p_{0k}^t = 1$ if $t = k$ and $= 0$ otherwise;

(c) $p_{jk}^0 = 0$ for $k \neq j$, and $p_{jj}^0 = n_j$, the total of every row of B_j;

(d) $\sum_j p_{jk}^t = n_k$;

(e) $n_k p_{sj}^k = n_j p_{sk}^j$.

Proofs:

(a) $B_k B_j = B_k' B_j' = (B_j B_k)' = (\sum_i p_{jk}^i B_i)' = \sum_i p_{jk}^i B_i' = \sum_i p_{jk} B_i = B_j B_k$, and hence (a).

(b) $B_k = B_0 B_k = \sum_i p_{0k}^t B_t$, hence (b).

(c) $\sum_{i=1}^{v} b_{tj}^i b_{tk}^i = 0$ for $j \neq k$ because of (4.2), and equals the row total of the tth row of B_j when $j = k$. Because of (4.5), p_{jj}^0 is the total of every row of B_j, say n_j, while $p_{jk}^0 = 0$ for $j \neq k$. From (4.2) it follows that $\sum_j n_j = v$.

(d) $\left(\sum\limits_{j} B_j\right) B_k = J_v B_k = n_k J_v = \sum\limits_{t} n_k B_t$ and also

$\sum\limits_{j}(B_j B_k) = \sum\limits_{j}\sum\limits_{t} p_{jk}^{t} B_t = \sum\limits_{t}\left(\sum\limits_{j} p_{j.}^{t}\right) B_t$.

Comparison of coefficients establishes (d).

(e) Consider the matrix products $M_j M_k$ and $M_k M_j$, where

$$M_j = \begin{pmatrix} p_{0j}^0 & \cdots & p_{0j}^m \\ - & - & - \\ p_{mj}^0 & \cdots & p_{mj}^m \end{pmatrix} .$$

Because of (a), the two matrix products are equal, and from this fact we deduce (4.6) and, in particular, (e).

Self-dual designs

Let a symmetric p.b.i.b.d. be given with incidence matrix A and with association matrices $B_i (i = 0, \ldots, m)$. The dual design will have the same parameters $v = b$ and $r = k$. In [54] A.J. Hoffman has derived sufficient (though not necessary) conditions for the dual to have also the parameters p_{jk}^i and λ_i.

We establish conditions which ensure that symmetric matrices, say C_i, with elements 0 and 1 exist, which satisfy (4.1), (4.2), (4.5) with the same parameters as the B_i, and satisfy also, instead of (4.7), the relation (4.7a) $A'A = \lambda_i C_i$.

We prove that the matrices $C_i = A^{-1} B_i A$ satisfy these conditions, provided the following conditions hold:

Condition I A is not singular (i.e. A^{-1} exists).

Condition II   The greatest common divisor of all determinants of order m contained in the $(m+2)$ by m matrix

$$\begin{pmatrix} 1 & \cdots & m \\ \sum\limits_{t} \lambda_t p_{0t}^1 & \cdots & \sum\limits_{t} \lambda_t p_{0t}^m \\ - & - & - \\ \sum\limits_{t} \lambda_t p_{mt}^1 & \cdots & \sum\limits_{t} \lambda_t p_{mt}^m \end{pmatrix}$$

is unity.

Proof of symmetry

$$AA'B_t = \sum\limits_{i} \lambda_i B_i B_t, \quad \text{and} \quad B_t AA' = \sum\limits_{i} \lambda_i B_t B_i .$$

Hence $AA'B_t = B_t AA'$, and $B_t A = AA'B_t A'^{-1}$. Therefore,

$$C_t = A^{-1} B_t A = A^{-1} AA' B_t A'^{-1} = C_t' .$$

Proof of (4.1) $C_0 = A^{-1} B_0 A = A^{-1} I_v A = I_v$.

Proof of (4.2) $\sum_i C_i = \sum_i A^{-1} B_i A = A^{-1} J_v A$.

We have to show that this equals J_v, and this is a consequence of $J_v A = A J_v$, which will now be proved.

We have $A J_v = r J_v$. Also $A' J_v = A^{-1} A A' J_v = A^{-1} \sum_i \lambda_i B_i J_v = c J_v$, where c is a constant, equal to the sum of the entries in any one column of A. Since the total of all entries of A is rv, it follows that $c = r$. But $A' J_v = r J_v$ as well as J_v are symmetric. Hence $J_v A = r J_v = A J_v$.

Proof of (4.5)

$$C_j C_k = (A^{-1} B_j A)(A^{-1} B_k A) = \sum_t A^{-1} p_{jk}^t B_t A = \sum_t p_{jk}^t C_t.$$

Proof of (4.7a) $A' A = A^{-1} A A' A = A^{-1} \sum_i \lambda_i B_i A = \sum_i \lambda_i C_i$.

The C_i will be association matrices if their elements are 0 or 1. Multiply (4.7) on the left by A^{-1} and on the right by $B_j A$. Then

$$A' B_j A = A' \sum_i \lambda_i B_i B_j A = \sum_i \sum_s \lambda_s p_{js}^i C_i \quad (j = 0, 1, ..., m).$$

Imagine these matrix equations written down explicitly for all elements of the C_i. Reference to a theorem in [122] proves that these elements are all integers, by virtue of Condition II. We show now that these elements are 0 or 1. The sum of the squares of all elements in C_i is equal to the trace (i.e. the sum of the diagonal elements) of $C_i C_i' = C_i^2$. By virtue of (4.5) this is also the trace of $\sum_t p_{ii}^t C_t$. The trace of C_t equals that of B_t, i.e. 0 for $t \neq 0$, and v for $t = 0$. Hence the sum of the squares of the elements of C_i is $n_i v$.

The sum of all elements of C_i is also $n_i v$, because all row (and column) totals are n_i. But the sum of integers can only be equal to the sum of their squares if all elements are either 0 or 1.

For $m = 1$ this theorem means that the dual of a symmetric b.i.b.d. is again a b.i.b.d. with the same parameters. This is a result which we have already obtained. (In this case the conditions are always satisfied, provided $r > \lambda$.)

The methods of block section and block intersection are applicable to p.b.i.b.d's as well, provided they are of linked block type. C.R. Nair has discussed cases where the association scheme of the new design equals that of the original one (see [69]).

p.b.i.b.d's through inversion

The dual of a b.i.b.d. with incidence matrix A is the b.i.b.d. whose incidence matrix is A'. The latter design is also said to have been obtained from the former by inversion.

If the parameters of the original b.i.b.d. are denoted by $v, b, k,$ and r, with their usual meaning, then the dual design has b varieties in v blocks of r varieties each, and each variety is repeated k times. Any pair of varieties will appear in the new design as often as there were common varieties in the blocks to which the two varieties correspond by their duality.

We shall now quote cases in which configurations turn, by inversion, into p.b.i.b.d's.

Case (1) ([101])

A b.i.b.d. with v varieties, $b = \binom{v}{k}$, $r = \binom{v-1}{k-1}$, $\lambda = \binom{v-2}{k-2}$ (an "unreduced" b.i.b.d.), turns into a p.b.i.b.d. unless $v = k+1$, or $k = 1$.

Let the varieties be $1, 2, \ldots, v,$ and take the block $B = (1, \ldots, k)$. There are $\binom{v-k}{k}$ blocks with no element in common with B. There are $\binom{v-k}{k-1}$ blocks with only the variety 1 in common with B, and $k \binom{v-k}{k-1}$ blocks with precisely one element in common with B. Similarly, there are $\binom{k}{i}\binom{v-k}{k-i}$ blocks with precisely i elements in common with B. Of course, $\sum_{i=0}^{k} \binom{k}{i}\binom{v-k}{k-i} = \binom{v}{k} = b$, as it must be. (We have included B itself in this count.) It will be noticed that $\binom{k}{i}\binom{v-k}{k-i} = 0$ when $i > k$ or when $i < 2k - v$.

Upon inversion there will be $\binom{k}{i}\binom{v-k}{k-i}$ varieties which appear i times in the same block with any other variety. If two varieties which appear $k-i$ times together are ith associates, then $\lambda_i = k-i$ and $n_i = \binom{k}{k-i}\binom{v-k}{i}$. (It is here that $v-k > 1$ is necessary; otherwise we again have a b.i.b.d.) These numbers are independent of which variety we started from, and so are the p_{jt}^i which we are now going to determine. (We use t as the second subscript, because k is used for the number of varieties within a block of the original b.i.b.d.)

Consider, in the original b.i.b.d., two blocks with $k-i$ varieties in common. Let these be $a_1, \ldots, a_{k-i}$, and let the remaining varieties in the first block be $b_1, \ldots, b_i$ and in the second block $c_1, \ldots, c_i$. Denote those varieties which are not within the sets just mentioned by $d_1, \ldots, d_{v-k-i}$. We want to find a block of k varieties with t from among the a's or the b's, and j from among the a's or the c's.

If s varieties are taken from among the a's, $k-t-s$ from among the b's, and $k-j-s$ from among the c's, then there remain $t+j+s-k$ to be taken from among the d's. Given s, this can be done in $\binom{k-i}{s} \times \binom{i}{k-t-s} \binom{i}{k-j-s} \binom{v-k-i}{t+j+s-k}$ different ways, and since s may be 0, or 1,...., or k (if we take too high a value, then the corresponding binomial coefficient will be zero), we have to sum the above product of four binomial coefficients over all those values of s to obtain p_{jt}^i.

Case (2) ([101] and [113])

If $\lambda = 1$, then the inversion of a b.i.b.d. produces a p.b.i.b.d., unless the original b.i.b.d. is symmetric, in which case its dual is also a b.i.b.d.

Take a block B and one variety in it, say x. There are $r-1$ more blocks in which x appears, and this is true whichever of the varieties in B we take. However, since $\lambda = 1$, any other variety will reappear in $r-1$ blocks different from those in which x reappears. Hence, if after inversion we call first associates those varieties which appear in the same block, and all other pairs second associates, then $n_1 = k(r-1)$, $n_2 = b - k(r-1) - 1$, $\lambda_1 = 1$, $\lambda_2 = 0$.

For a symmetric b.i.b.d. we have $k(r-1) = r(k-1) = v-1$, so that $n_1 = v-1$, i.e. all varieties are first associates.

To find the parameters of the second kind, we determine first p_{11}^2. From the original blocks let us take two with no variety in common, say the blocks $\mathbf{a} = (a_1, \ldots, a_k)$ and $\mathbf{b} = (b_1, \ldots, b_k)$. There is precisely one block which contains a_i as well as b_j, and this block has precisely one element in common with block $\mathbf{a}$, and also with block $\mathbf{b}$. The subscripts i and j take, independently, k different values. Therefore $p_{11}^2 = k^2$, and from this we obtain, through the relations which connect the parameters,

$$(p_{jt}^1) = \begin{pmatrix} r-2+(k-1)^2 & (r-k)(k-1) \\ (r-k)(k-1) & b+r-k(2r-k)-1 \end{pmatrix}$$

$$(p_{jt}^2) = \begin{pmatrix} k^2 & k(r-k-1) \\ k(r-k-1) & b+k^2-2k(r-1)-2 \end{pmatrix}$$

In this derivation we have made use of the relations $b = vr/k$ and $r(k-1) + 1 = v$.

If $k=2$, then we have again a special case of (1). In [101] P.M. Roy has constructed more p.b.i.b.d's by inverting b.i.b.d's. In [113] S.S. Shrikhande has studied p.b.i.b.d's which are duals of b.i.b.d's with $\lambda = 1$ obtained from systems of difference sets.

Case (3) ([101] and [113]).

The inversion of a b.i.b.d. with $\lambda = 2$ which was obtained from a symmetric b.i.b.d. by block section leads to a p.b.i.b.d. with parameters $n_1 = \binom{k}{2}$, $n_2 = 2k$, $\lambda_1 = 2$, $\lambda_2 = 1$, and

$$(p^1_{jk}) = \begin{pmatrix} \binom{k-2}{2} & 2(k-2) \\ 2(k-2) & 4 \end{pmatrix}, \quad (p^2_{jk}) = \begin{pmatrix} \binom{k-1}{2} & k-1 \\ k-1 & k \end{pmatrix}.$$

For the proof, we refer the reader to the original sources.

Case (4) ([113])

The dual of a b.i.b.d. with $v = \binom{r-1}{2}$, $b = \binom{r}{2}$, $k = r-2$, $\lambda = 2$ is a p.b.i.b.d. with $n_1 = 2(r-2)$, $n_2 = \binom{r-2}{2}$, $\lambda_1 = 1$, $\lambda_2 = 2$.

$$(p^1_{jk}) = \begin{pmatrix} r-2 & r-3 \\ r-3 & \binom{r-3}{2} \end{pmatrix}, \quad (p^2_{jk}) = \begin{pmatrix} 4 & 2(r-4) \\ 2(r-4) & \binom{r-4}{2} \end{pmatrix}$$

The proof depends on a theorem in [58]. We refer the reader to the source given above. S.S. Shrikhande has shown in [119] that the p.b.i.b.d. thus obtained is always of triangular type (see Chapter V).

Case (5)

Consider an affine resolvable b.i.b.d. with $v = n^2m$, $b = nr$, $k = nm$. After inversion we have a p.b.i.b.d. with $n_1 = b-n$, $n_2 = n-1$, $\lambda_1 = m$, $\lambda_2 = 0$,

$$(p^1_{jk}) = \begin{pmatrix} b-2n & n-1 \\ n-1 & 0 \end{pmatrix}, \quad (p^2_{jk}) = \begin{pmatrix} b-n & 0 \\ 0 & n-2 \end{pmatrix}.$$

Case (6) ([116])

If from the affine resolvable configuration of case (5) we omit all blocks that contain a particular variety, then we obtain a symmetric configuration with $n^2m - 1 = (n-1)r$ varieties and blocks. The blocks contain $nm = r - \lambda$ varieties each, and each variety is repeated nm times. The dual of this design is a p.b.i.b.d. with $\lambda_1 = 0$, $\lambda_2 = m$, $n_1 = n-2$, $n_2 = n^2m - n$. A further examination of duality and partial balance is contained in [94].

Examples for (5) and (6)

The design

$$
\begin{array}{cccc}
1\,2\,3 & 1\,4\,5 & 1\,6\,7 & 1\,8\,9 \\
4\,6\,8 & 2\,6\,9 & 2\,5\,8 & 2\,4\,7 \\
5\,7\,9 & 3\,7\,8 & 3\,4\,9 & 3\,5\,6
\end{array}
$$

is affine resolvable. Inverting, we have

$$
\begin{array}{ccc}
1\,4\,7\,10 & 2\,4\,9\,11 & 3\,6\,7\,11 \\
1\,5\,8\,11 & 3\,4\,8\,12 & 2\,6\,8\,10 \\
1\,6\,9\,12 & 2\,5\,7\,12 & 3\,5\,9\,11
\end{array}
$$

$n_1 = 12 - 3 = 9$, $n_2 = 3 - 1 = 2$, $\lambda_1 = 1$, $\lambda_2 = 0$,

$$
(p^1_{jk}) = \begin{pmatrix} 6 & 2 \\ 2 & 0 \end{pmatrix}, \quad (p^2_{jk}) = \begin{pmatrix} 9 & 0 \\ 0 & 1 \end{pmatrix}.
$$

Omit all blocks with 9, say; we obtain

$$
\begin{array}{cccc}
1\,2\,3 & 1\,6\,7 & 1\,4\,5 & 2\,4\,7 \\
4\,6\,8 & 2\,5\,8 & 3\,7\,8 & 3\,5\,6
\end{array}
$$

Inverting, we have, with simplified notation,

$$
\begin{array}{cccccccc}
1\,3\,5 & 2\,3\,7 & 4\,5\,7 & 1\,6\,7 & 3\,6\,8 & 2\,4\,6 & 1\,4\,8 & 2\,5\,8
\end{array}
$$

$$
(p^1_{jk}) = \begin{pmatrix} 0 & 0 \\ 0 & 6 \end{pmatrix}, \quad (p^2_{jk}) = \begin{pmatrix} 0 & 1 \\ 1 & 4 \end{pmatrix}.
$$

It should be noted that in all the cases (1) to (6) the parameters v, b, r, and k refer to the original, and not to the inverted design.

EXERCISES

(1) Construct a design from the simple rectangular lattice for $n = 4$.
(2) Determine the parameters of the p.b.i.b.d. derived from the difference set $1, 2, 3 \pmod 5$ (see page 58).

(3) Find the association matrix of the scheme of the following design, and determine AA' and all $B_i B_j$.

$$3478, \quad 1234, \quad 2367, \quad 5678, \quad 1256, \quad 1458.$$

$$\lambda_0 = 3, \ \lambda_1 = 2, \ \lambda_2 = 1, \ \lambda_3 = 0, \ n_1 = n_2 = 3, \ n_3 = 1.$$

(4) Invert

		12	23		123	145	167	189
(i)		13	24	(ii)	468	269	258	247
		14	34		579	378	349	356

(5) Construct a p.b.i.b.d. by inverting the design obtained by block section from

12345	1569E	34689
12678	236XE	3578E
1379X	2479E	4567X
148XE	2589X	

$$v = b = 11, \quad r = k = 5, \quad \lambda = 2.$$

(Example from [54].)

(6) Find the two triple rectangular lattices which are p.b.i.b.d's (see page 60) and determine their association schemes. How would you describe them as configurations? Invert the first p.b.i.b.d.

CHAPTER V

PARTIALLY BALANCED INCOMPLETE BLOCK DESIGNS
WITH TWO ASSOCIATE CLASSES

Introduction

To begin with, we repeat for convenience the formulae for the parameters of a p.b.i.b.d. as they apply to designs with two associate classes:

$$vb = rk, \quad n_1 + n_2 = v - 1, \quad \lambda_1 n_1 + \lambda_2 n_2 = r(k-1),$$

$$p_{11}^1 + p_{12}^1 = n_1 - 1, \quad p_{11}^2 + p_{12}^2 = n_1, \quad p_{12}^1 + p_{22}^1 = n_2, \quad p_{12}^2 + p_{22}^2 = n_2 - 1$$

$$n_1 p_{22}^1 = n_2 p_{12}^2 \qquad n_1 p_{12}^1 = n_2 p_{11}^2$$

When the parameters n_1 and n_2 are known, then only one of the p_{jk}^i remains independent.

The definition of the p_{jk}^i demands that they should be independent of which pair of ith associates we start from. In [12] Bose and Clatworthy have pointed out that for designs with two associates it is sufficient to demand this independence only for p_{11}^1 and p_{11}^2; that of the others will then follow. If x and y are any two varieties which are first associates, then let $p_{jk}^1(x,y)$ be the number of jth associates of x and kth associates of y. We have then

$$p_{11}^1(x,y) + p_{12}^1(x,y) = n_1 - 1,$$
$$p_{11}^1(x,y) + p_{21}^1(x,y) = n_1 - 1,$$

and
$$p_{21}^1(x,y) + p_{22}^1(x,y) = n_2.$$

It follows that if $p_{11}^1(x,y)$ is independent of x and y, then this is also true of all the other p_{jk}^1, and the same follows for $p_{jk}^2(x,y)$ in a similar way.

On the other hand, it is possible that $p_{jk}^2(x,y)$ is independent of the pair x and y, but $p_{jk}^1(x,y)$ is not. The following example is given in [12]:

01	12	23	34	45	56	60
03	14	25	36	40	51	62

This proves the statement, if first associates are varieties which appear in the same block. In this example $p_{11}^1(0,1) = 1$, for instance, but $p_{11}^1(0,3) = 2$.

A design with equal block sizes (k) and an equal number of replications (r) of all varieties is called a (b_k, v_r) configuration. If it is symmetric, i.e. $b = v$, and $r = k$, then we denote it by (v_r) (see [129]).

Let varieties in the same block be first, and other pairs second associates. Then $\lambda_1 = 1$ and $\lambda_2 = 0$.

Examples A (7_3) is the $PG(2,2)$, a b.i.b.d. The (8_3)

$$
\begin{array}{cccc}
124 & 167 & 235 & 457 \\
138 & 278 & 346 & 568
\end{array}
$$

is a p.b.i.b.d. with parameters $v = b = 8$, $r = k = 3$, $n_1 = 6$, $n_2 = 1$ and $(p_{jk}^1) = \begin{pmatrix} 4 & 1 \\ 1 & 0 \end{pmatrix}$, $(p_{jk}^2) = \begin{pmatrix} 6 & 0 \\ 0 & 0 \end{pmatrix}$.

There exist three different (9_3), with $v = b = 9$, $r = k = 3$, $n_1 = 6$, $n_2 = 2$ and the following values of

(p_{jk}^1)	and	(p_{jk}^2)		for the configuration			
4 1		5 1	123	145	189	248	269
1 1		1 0	356	379	467	578	
3 2		5 1	123	145	168	257	289
2 0		1 0	356	349	478	679	
3 2		6 0	123	148	159	247	269
2 0		0 1	357	368	456	789	

(This is the configuration illustrating the theorem of Pappus)

The Desargues configuration illustrates a (10_3) which is a p.b.i.b.d. with $v = b = 10$, $r = k = 3$, $n_1 = 6$, $n_2 = 3$, $(p_{jk}^1) = \begin{pmatrix} 3 & 2 \\ 2 & 1 \end{pmatrix}$, $(p_{jk}^2) = \begin{pmatrix} 4 & 2 \\ 2 & 0 \end{pmatrix}$.

Other p.b.i.b.d's can be obtained from the points and faces of polyhedra.

A design with either $\lambda_1 = 0$ or $\lambda_2 = 0$ is called *simple* in [13]. A thorough examination of simple designs with $k > r \geqslant 2$ is contained in [12]. See also [8] for $k > r = 2$. An enumeration of p.b.i.b.d's with two associate classes and $r = 3$ is given in [99].

If in a projective geometry we take all flats of a given dimension as blocks, and all the points on it as varieties, then we obtain a b.i.b.d., as we have seen. If we omit suitably chosen blocks or varieties, and call first associates those varieties which appear together in one of the remaining blocks, while any other pair of points are second associates, then we obtain simple p.b.i.b.d's. We shall give a few examples of this procedure.

Omit one point, say P, of a $PG(m,s)$ and all t-flats through it, but no further points in the latter. There remain $v = s(s^m - 1)/(s-1)$ varieties.

Originally we had $V(m,t;s)$ different t-flats. Of these we have omitted all those through P, i.e. $V(m-1, t-1; s)$, so that there remain

$$
\begin{aligned}
b &= V(m-1, t-1; s)(s^{m+1} - s^{t+1})/(s^{t+1} - 1) \\
&= s^{t+1} \frac{(s^m - 1) \ldots (s^{m-t+1} - 1)}{(s^t - 1) \ldots (s-1)} \cdot \frac{(s^{m-t} - 1)}{(s^{t+1} - 1)} .
\end{aligned}
$$

Every one of these blocks contains $k = (s^{t+1} - 1)/(s-1)$ varieties. It follows that the number of times each variety appears is $r = bk/v =$

$$
\frac{(s^m - s^t)}{(s^t - 1)} \cdot \frac{(s^{m-1} - 1) \ldots (s^{m-t+1} - 1)}{(s^{t-1} - 1) \ldots (s-1)} .
$$

Two varieties will appear in the same t-flat unless their line passes through P: then and then only will all the t-flats have been omitted in which these appeared together in the original $PG(m,s)$. Thus $n_2 = s-1$ (all points on the line through P and through a chosen point, excluding these two) and hence $n_1 = s^2(s^{m-1} - 1)/(s-1)$. λ_1 is the number of t-flats through a straight line, less those through this line and P, i.e.

$$
\begin{aligned}
\lambda_1 &= V(m-2, t-2; s) - V(m-3, t-3; s) \\
&= \frac{(s^{m-1} - s^{t-1})}{(s^{t-1} - 1)} \cdot \frac{(s^{m-2} - 1) \ldots (s^{m-t+1} - 1)}{(s^{t-2} - 1) \ldots (s-1)} .
\end{aligned}
$$

We turn now to the parameters of the second kind. We have $p_{22}^1 = 0$, because a variety's second associates are all in the line through its corresponding point and P, and only two points which are on a line with P can have a common second associate. From this the other values follow:

$$(p^1_{jk}) = \begin{pmatrix} s^m + \ldots + s^2 - s & s-1 \\ s-1 & 0 \end{pmatrix}$$

$$(p^2_{jk}) = \begin{pmatrix} s^m + \ldots + s^2 & 0 \\ 0 & s-2 \end{pmatrix}.$$

We notice that these latter parameters are independent of t. If, in particular, $t = m-1$, then the expressions reduce to $v = s(s^m - 1)/(s-1)$, $b = s^m$, $k = (s^m-1)/(s-1)$, $r = s^{m-1}$, $n_1 = s^2(s^{m-1}-1)/(s-1)$, $n_2 = s-1$, $\lambda_1 = s^{m-2}$, $\lambda_2 = 0$. If $m = 2$, $t = 1$, then $v = s^2 + s$, $b = s^2$, $k = s+1$, $r = s$, $n_1 = s^2$, $n_2 = s-1$, $\lambda_1 = 1$, $\lambda_2 = 0$.

More generally, we could start from any b.i.b.d. with parameters $v, b, r, k, \lambda = 1$ and omit one variety and all blocks that contain it. We obtain then a p.b.i.b.d. with $v-1$ varieties, $b-r$ blocks, $r-1$ replications and $\lambda_1 = 1$, $\lambda_2 = 0$.

An analogous construction can start from a $EG(m,s)$.

We omit one point of the $EG(m,s)$ and all the t-flats through it. There remain $v = s^m - 1$ points. Originally we had $V(m-1, t-1; s) \cdot s^{m-t}$ different t-flats. Of these we have omitted all those through a chosen point, $V(m-1, t-1; s)$ in number, so that there remain $b = V(m-1, t-1; s) \times (s^{m-t} - 1)$ t-flats. All of them contain $k = s^t$ points, hence $r = bk/v = V(m-2, t-1; s)s^t$. By the same argument as before, we now obtain $n_2 = s-2$, and hence $n_1 = s^m - s$. Also we have $\lambda_1 = r(k-1)/n_1 = V(m-3, t-2; s)s^{t-1}$, unless $m = 2$, when $\lambda_1 = 1$. Again, $p^1_{22} = 0$, so that

$$(p^1_{jk}) = \begin{pmatrix} s^m - 2s+1 & s-2 \\ s-2 & 0 \end{pmatrix}, \quad (p^2_{jk}) = \begin{pmatrix} s^m - s & 0 \\ 0 & s-3 \end{pmatrix}.$$

(When $s = 2$, then only first associates exist, and we have a b.i.b.d.) In particular, when $t = m-1$, then $v = b = s^m - 1$, $r = k = s^{m-1}$, $\lambda = s^{m-2}$.

A p.b.i.b.d. where p^1_{22}, p^2_{12} and p^2_{21} are not all zero can be derived from the correspondence between a point $(y_0, \ldots, y_m)$ and the hyperplane $\sum_i (\sum_j a_{ij} y_j) x_i = 0$ with skew-symmetric matrix (a_{ij}). Let m be odd, and let the points of a $PG(m,s)$ be varieties, and let blocks contain the varieties of those points on a line in a hyperplane passing through the point corresponding to the latter. There are then $v = (s^{m+1} - 1)/(s-1)$ varieties. On every line there are $k = s+1$ points, and through each point there are $r = (s^{m-1} - 1)/(s-1)$ lines of the hyperplane; this is the number of replications of the variety corresponding to that point. Hence the number of blocks is

$$b = vr/k = [(s^{m+1}-1)(s^{m-1}-1)]/[(s-1)^2(s+1)].$$

Let first associates be varieties in the same block, and let any other pair be second associates. We then have a p.b.i.b.d. with $n_1 = s(s^{m-1}-1)/(s-1)$ (i.e. s times the number of lines through a point in the hyperplane), $n_2 = (s^{m+1}-1)/(s-1) - (s^m-s)/(s-1) - 1 = s^m$, $\lambda_1 = 1$, and $\lambda_2 = 0$.

The parameters of the second kind are determined as follows: two points which are first associates lie each on the hyperplane corresponding to one of them, and on that corresponding to the other. Hence those other varieties which are first associates of both of the former ones lie in the intersection of the two hyperplanes. Excluding the two points themselves, we have thus $p_{11}^1 = (s^{m-1}-1)/(s-1) - 2$, and hence

$$(p_{jk}^1) = \begin{pmatrix} \dfrac{(s^{m-1}-2s+1)}{(s-1)} & s^{m-1} \\ s^{m-1} & s^{m-1}(s-1) \end{pmatrix}, \quad (p_{jk}^2) = \begin{pmatrix} \dfrac{s^{m-1}-1}{s-1} & s^{m-1}-1 \\ s^{m-1}-1 & s^{m-1}(s-1) \end{pmatrix}.$$

This construction is a fairly obvious extension of that given by Clatworthy, in [34] for $m=3$. For $m=3$, $s=2$ the p.b.i.b.d. is mentioned in [19]. It is of triangular type, see later, page 84.

A further generalisation, taking flats of higher dimension instead of the lines used here for blocks, is contained in [1]. The values of p_{jk}^1 and p_{jk}^2 are again independent of the dimension of the flats.

We mention one more construction of a p.b.i.b.d. from a $PG(m,s)$. Choose a point P, and t of the $(s^m-1)/(s-1)$ lines through it. Let the other s points on these lines correspond to varieties, so that there are $v = s.t$ varieties altogether. Let the $b = \dfrac{s^{m+1}-1}{s-1} - \dfrac{s^m-1}{s-1} = s^m$ hyperplanes not passing through P correspond to blocks. Then each point that corresponds to a variety will be contained in $r = \dfrac{s^m-1}{s-1} - \dfrac{s^{m-1}-1}{s-1} = s^{m-1}$ hyperplanes representing blocks, and each block will contain $k=t$ varieties. We call "first associates" points on one of the t chosen lines, and any other pair "second associates", so that $n_1 = s-1$ and $n_2 = s(t-1)$. Also $\lambda_1 = 0$ and $\lambda_2 = s^{m-2}$. We find

$$(p_{jk}^1) = \begin{pmatrix} s-2 & 0 \\ 0 & s(t-1) \end{pmatrix}, \quad (p_{jk}^2) = \begin{pmatrix} 0 & s-1 \\ s-1 & s(t-2) \end{pmatrix}.$$

Another method of obtaining families of simple p.b.i.b.d's was described in [109] by E. Seiden.

We have seen that in a $PG(2,2^n)$ there exist sets of 2^n+2 points each (a conic and its nucleus), no three of which are collinear. There are $\binom{2^n+2}{2} = (2^{n-1}+1)(2^n+1)$ lines through two points of such a set, and no lines with precisely one, hence there are $(2^{2n}+2^n+1) - (2^{n-1}+1)(2^n+1) = 2^{2n-1}-2^{n-1}$ lines with no point of the set on any of them.

We take as varieties the $2^{2n}-1$ points outside the chosen set and as blocks the lines containing two points of the set (each point lies on $2^{n-1}+1$ of these lines). This gives us one family of p.b.i.b.d's. In a second design we take as blocks the lines containing no point of the set (each point lies on 2^{n-1} of these).

Let two varieties be first associates if they appear in the same block, and second associates if they do not. Then $\lambda_1 = 1$ and $\lambda_2 = 0$. The other parameters are found to be

1st design		2nd design
$2^{2n}-1$	v	$2^{2n}-1$
$(2^{n-1}+1)(2^n+1)$	b	$2^{n-1}(2^n-1)$
$2^{n-1}+1$	r	2^{n-1}
2^n-1	k	2^n+1
$2^{2n-1}-2$	n_1	2^{2n-1}
2^{2n-1}	n_2	$2^{2n-1}-2$
$\begin{pmatrix} 2^{2n-2}-3 & 2^{2n-2} \\ 2^{2n-2} & 2^{2n-2} \end{pmatrix}$	(p_{jk}^1)	$\begin{pmatrix} 2^{2n-2} & 2^{2n-2}-1 \\ 2^{2n-2}-1 & 2^{2n-2}-1 \end{pmatrix}$
$\begin{pmatrix} 2^{2n-2}-1 & 2^{2n-2}-1 \\ 2^{2n-2}-1 & 2^{2n-2} \end{pmatrix}$	(p_{jk}^2)	$\begin{pmatrix} 2^{2n-2} & 2^{2n-2} \\ 2^{2n-2} & 2^{2n-2}-3 \end{pmatrix}$

In [95] p.b.i.b.d's are constructed from the geometry of quadrics in spaces of higher dimension.

Shrikhande and Singh have pointed out in [121] that if in a p.b.i.b.d. with two associate classes $p_{11}^1 = p_{11}^2$, then a symmetric b.i.b.d. can be constructed by forming v blocks from the first associates of each variety. We obtain a b.i.b.d. of linked block type with $k = n_1 = r$.

Example

We start from a b.i.b.d. with $\lambda = 1$, $r = 2k+1$, and hence $v = k(2k-1)$,

$b = 4k^2 - 1$, and take its dual (see case (2) on page 67, obtaining a p.b.i.b.d. with $v = 4k^2 - 1$, $n_1 = 2k^2$, $p_{11}^1 = k^2$, $p_{11}^2 = k^2$. If we then form a symmetric b.i.b.d. by the method just described, we obtain a Hadamard configuration (see page 14) with $v = 4k^2 - 1$, $r = 2k^2$, $\lambda = k^2$.

In [10] R.C. Bose has applied graph-theoretic methods to the study of association schemes of p.b.i.b.d's with two associate classes.

CLASSIFICATION

In [19] the authors have classified p.b.i.b.d's with two associate classes according to their association schemes. We shall now consider this topic.

(A) Group-divisible designs

We start with the group-divisible (GD) scheme and shall call a p.b.i.b.d. with such a scheme a GD-design.

Let $v = m.n$ and write the varieties into m columns (called groups) and n rows. Let two variables in the same column be first, and any other pair second associates. Then $n_1 = n - 1$, $n_2 = n(m-1)$. Also $p_{12}^1 = 0$, because the relation of being first associates is transitive (as in the rectangular scheme, of which this is a modification, already mentioned in Chapter IV). From this, or just as easily from first principles, we obtain

$$(p_{jk}^1) = \begin{pmatrix} n-2 & 0 \\ 0 & n(m-1) \end{pmatrix}, \quad (p_{jk}^2) = \begin{pmatrix} 0 & n-1 \\ n-1 & n(m-2) \end{pmatrix}.$$

We have already met this design, derived in a different way, on page 75. Conversely, if a p.b.i.b.d. with two associate classes has $p_{12}^1 = 0$, then it must be a GD-design. We have then $p_{11}^1 = n_1 - 1$, so that a first associate of any variety is also a first associate of all its first associates. Hence the v varieties can be divided into groups, each of $n_1 + 1 = n$, say, varieties, and those in the same group are first associates.

If in a design $p_{12}^2 = 0$, we again have a GD-design, but have to call all varieties in the same group second associates.

A fairly obvious way of constructing GD-designs is to start from a b.i.b.d. with parameters v,b,r,k,λ and to replace each variety by n new ones. We have then altogether $v.n$ new varieties, and if any two of these which have replaced the same original variety are called first, and any other pair second associates, then we have a GD-design with $\lambda_1 = r$, $\lambda_2 = \lambda$, and $m = v$ (see Exercise V.5).

If in a GD-design $\lambda_2 = 0$, then it consists of disconnected groups of blocks, i.e. of m isomorphic b.i.b.d's each of n varieties.

Families of symmetric GD-designs are given in [35] as follows:

(1) $b = v = 2n$ blocks $(a_1, \ldots, a_n, b_i)$ and $(b_1, \ldots, b_n, a_i)$,
   $(i = 1, \ldots, n)$ with the association scheme

$$
\begin{array}{cc}
a_1 & b_1 \\
a_2 & b_2 \\
- & - \\
- & - \\
a_n & b_n
\end{array}
$$

Then $m = 2$, $r = k = n+1$, $n_1 = n-1$, $n_2 = n$, $\lambda_1 = n$, $\lambda_2 = 2$.

(2) $b = v = 3n$ blocks $(a_1, \ldots, a_n, b_i)$, $(b_1, \ldots, b_n, c_i)$, $(c_1, \ldots, c_n, a_i)$
   $(i = 1, \ldots, n)$ with the association scheme

$$
\begin{array}{ccc}
a_1 & b_1 & c_1 \\
- & - & - \\
- & - & - \\
a_n & b_n & c_n
\end{array}
$$

Then $m = 3$, $r = k = n+1$, $n_1 = n-1$, $n_2 = 2n$, $\lambda_1 = n$, $\lambda_2 = 1$.

Symmetric GD-designs can also be constructed from a set $d_1, \ldots, d_k$ of integers mod $v = m.n$ such that among the $k(k-1)$ differences every integer congruent to multiples of m appears λ_1 times and all others λ_2 times. (As regards a necessary condition for the existence of such sets when $n \equiv 3$ (mod 4) see [129], page 22.) The ith column of the association scheme is then $(i, m+i, 2m+i, \ldots, (n-1)m+i)$, and the blocks of the design are $(d_1+t, \ldots, d_k+t)$ $t = 0, \ldots, v-1$.

In [124] D.A. Sprott constructs a GD-design as follows. Consider a b.i.b.d. with $v = b = s^4 + s^2 + 1$, $r = k = s^2 + 1$, $\lambda = 1$, represented, for instance, by the lines of a $PG(2, s^2)$. We omit the $s^2 + s + 1$ blocks given by the lines of a $PG(2, s)$ contained in the $PG(2, s^2)$, and the $s^2 + s + 1$ varieties corresponding to the points of the $PG(2, s)$. We then obtain a GD-design with $v = b = s^4 - s$, $k = r = s^2$, $m = s^2 + s + 1$, and $n = s^2 - s$.

Example

$(s=2)$. Association scheme

1	3	5	7	9	11	13
2	4	6	8	10	12	14

$n_1=1$, $n_2=12$, $v=b=14$, $r=k=4$, $\lambda_1=0$, $\lambda_2=1$.

2	3	5	9		1	4	6	10
4	5	7	11		3	6	8	12
6	7	9	13		5	8	10	14
1	8	9	11		2	7	10	12
3	10	11	13		4	9	12	14
1	5	12	13		2	6	11	14
1	3	7	14		2	4	8	13

In [19] we find this example, and another non-isomorphic GD-design with the same association scheme, namely $v=14$, $b=8$, $r=4$, $k=7$, $\lambda_1=0$, $\lambda_2=2$.

1	3	5	7	9	11	13
2	4	6	8	9	11	13
1	3	6	8	10	12	13
2	4	5	7	10	12	13
1	4	5	8	9	12	14
2	3	6	7	9	12	14
1	4	6	7	10	11	14
2	3	5	8	10	11	14

We have seen in Chapter IV that the dual of a b.i.b.d. with $\lambda=1$ is always a p.b.i.b.d. (see case (2) on page 67). If we start, in particular, with s^2 varieties, s^2+s blocks, s varieties in each block, and $s+1$ replications of each variety (lines in a $EG(2,s)$), then inversion produces a p.b.i.b.d. with $v=s^2+s$, $b=s^2$, $r=s$, $k=s+1$, $\lambda_1=1$, $\lambda_2=0$, $n_1=s^2$, $n_2=s-1$,

$$(p^1_{jk}) = \begin{pmatrix} s^2-s, & s-1 \\ s-1 & 0 \end{pmatrix}, \quad (p^2_{jk}) = \begin{pmatrix} s^2 & 0 \\ 0 & s-2 \end{pmatrix}.$$

This is a GD-design with $m=s+1$, $n=s$ where the role of the first and the second associates have been reversed. It is, incidentally, of linked block type.

The design obtained from an affine resolvable b.i.b.d. by inversion (see Chapter IV, case (5), page 68) is a GD-design if we exchange the names of first and second associates. That obtained in case (6) is a symmetric GD-design.

If in an orthogonal array $[\lambda n^2, m,n,2]$ we replace an element x in the ith row by $x+(i-1)n$, and take the columns as blocks, and the elements of the same row as belonging to the same group of the association scheme, then we have a GD-design with $v=mn$, $b=\lambda_2 n^2$, $r=\lambda_2 n$, $k=m$, $\lambda_1=0$, $\lambda_2=\lambda$.

If a completely orthogonal set of Latin cubes of the first order, of side s, exists, then a GD-design with $v=s^3$, $m=s$, $n=s^2$, $b=s^2(s+1)$, $r=s(s+1)$, $k=ns^2$, $\lambda_1=s$, $\lambda_2=s+1$, also exists. For the proof, write the varieties of the ith group $(i=1,\ldots,s)$ as a square of side s. We have then s^2 varieties in s squares. Superimpose the s^2+s-2 cubes on this arrangement. Picking up the treatments coinciding with the several elements of the 1st, 2nd, $\ldots$, cube and writing them in different blocks, we get $s(s^2+s-2)$ blocks, each of s^2 varieties. Add $2s$ further blocks, formed by writing into the jth block the varieties in the jth row of all the squares, and into the $(s+k)$th block the varieties in all the kth columns. The design obtained is of GD type, with the parameters as quoted above.

Consider now the incidence matrix of a GD-design, denoted by A. Then

$$AA' = \begin{pmatrix} E & F & \ldots & F \\ F & E & \ldots & F \\ & - & - & - \\ F & F & \ldots & E \end{pmatrix} \quad \text{where} \quad E = \begin{pmatrix} r & \lambda_1 & \ldots & \lambda_1 \\ \lambda_1 & r & \ldots & \lambda_1 \\ & - & - & - \\ \lambda_1 & \lambda_1 & \ldots & r \end{pmatrix}$$

and $F = \lambda_2 J_n$. Both E and F are n by n matrices.

To evaluate the determinant $|AA'|$, add the 2nd,3rd,$\ldots$, mnth row to the first. The latter row has then every element equal to rk. We take this factor out and subtract the first row λ_2 times from all other rows. Then we have

$$|AA'| = rk \begin{vmatrix} 1 & 1 & \ldots & 1 \\ \lambda_1-\lambda_2 & r-\lambda_2 & \ldots & \lambda_1-\lambda_2 \\ \lambda_1-\lambda_2 & \lambda_1-\lambda_2 & \ldots & r-\lambda_2 \end{vmatrix} \cdot \begin{vmatrix} r-\lambda_2 & \lambda_1-\lambda_2 & \ldots & \lambda_1-\lambda_2 \\ \lambda_1-\lambda_2 & r-\lambda_2 & \ldots & \lambda_1-\lambda_2 \\ \lambda_1-\lambda_2 & \lambda_1-\lambda_2 & \ldots & r-\lambda_2 \end{vmatrix}^{m-1}$$

The column sums of the second matrix are all $(r-\lambda_2)+(n-1)(\lambda_1-\lambda_2)$, and taking this factor out, we are left with the first matrix. Thus:

$$|AA'| = rk \begin{vmatrix} 1 & 1 & \cdots & 1 \\ \lambda_1 - \lambda_2 & r - \lambda_2 & \cdots & \lambda_1 - \lambda_2 \\ & -\ -\ - & \\ \lambda_1 - \lambda_2 & \lambda_1 - \lambda_2 & \cdots & r - \lambda_2 \end{vmatrix}^m [(r - \lambda_2) + (n-1)(\lambda_1 - \lambda_2)]^{m-1}$$

This equals $rk(r - \lambda_1)^{m(n-1)}(rk - v\lambda_2)^{m-1}$, as can be seen by subtracting the first column from the others in the matrix, and by noting that
$r - \lambda_2 + (n-1)(\lambda_1 - \lambda_2) = r + \lambda_1 n_1 - (1 - n_1)\lambda_2 = r + \lambda_1 n_1 + \lambda_2 n_2 - \lambda_2(1 + n_1 + n_2) = rk - v\lambda_2$.

Types of group-divisible designs

GD-designs are classified according to the sign of the factors of $|AA'|$ (see [14]). rk is, of course, always positive. $r - \lambda_1$ cannot be negative. If $r = \lambda_1$, then we call the design *singular*. In fact, any p.b.i.b.d. with two associate classes and $r = \lambda_1$ must be a GD-design, because in this case all first associates of a variety appear in a block if and only if that variety appears in the same block. Hence the relation of being first associate is transitive, $p_{12}^1 = 0$ (and $p_{11}^2 = 0$), and the design must be a GD-design (see [14]).

In a singular GD-design all varieties that are first associates of one another can be telescoped into a single variety: we then obtain a b.i.b.d. If the parameters of the original p.b.i.b.d. were $v = m.n$, b, $r = \lambda_1$, k, and λ_2, then the resulting b.i.b.d. will have $v' = m$ varieties, $r' = r$ replications, $b' = b$ blocks of $k' = k/n$ varieties each, and any pair of varieties will appear $\lambda' = \lambda_2$ times in the same block. It follows from $b' \geqslant v'$ that in a singular GD-design we have $b \geqslant m$.

Furthermore, in such a design we must have $rk - v\lambda_2 \geqslant 0$, because the left-hand side equals $(r'k' - v'\lambda')n$, and this cannot be negative.

If $r > \lambda_1$, then again $rk - v\lambda_2 \geqslant 0$. We can see this as follows.

Consider the submatrix A_1, consisting of the first $2n$ rows of A, i.e. of those referring to the varieties of the first two groups. Then

$$|A_1 A_1'| = \begin{vmatrix} E & F \\ F & E \end{vmatrix} = [r + (n-1)\lambda_1 + n\lambda_2](r - \lambda_1)^{2(n-1)}(rk - v\lambda_2).$$

But $|A_1 A_1'|$ must be non-negative, because A_1 has at least as many columns as it has rows* (see footnote on following page).

Since the expression in square brackets is positive, $rk - v\lambda_2 \geqslant 0$

follows. A *GD*-design with $r - \lambda_1 > 0$, $rk - v\lambda_2 > 0$ (and hence $|AA'| > 0$) is called *regular*, and one with $r - \lambda_1 > 0$ and $rk - v\lambda_2 = 0$ (and hence $|AA'| = 0$ as for a singular design) is called *semi-regular*.

For regular designs it follows, as for b.i.b.d's, that $b \geqslant v$, since the rank of AA', namely v, cannot be higher than that of A. If a regular *GD*-design is resolvable, then we can again derive $b \geqslant v + r - 1$, as on page 13.

Take those columns of A which correspond to the blocks of one complete replication. The entries of each row add up to 1, and therefore not more than $b - r + 1$ columns can be independent. Hence $b - r + 1 \geqslant v$, or $b \geqslant v + r - 1$.

Structure

In [37] Connor chooses $t < b$ blocks, and calls their incidence matrix A_0. He then evaluates the determinant

$$\begin{vmatrix} AA' & A_0 \\ A_0' & I_t \end{vmatrix} = (rk)^{1-t}(r-\lambda_1)^{v-t-m}(rk-v\lambda_2)^{m-t-1}|C_t|$$

and shows that, for a regular *GD*-design, $|C_t| \geqslant 0$ if $b > t + v$, $|C_t| = 0$ if $b < t + v$, while, if $b = t + v$, then

$$r^{-2(t-1)}(r-\lambda_1)^{v-t-m}(r^2-v\lambda_2)^{m-t-1}|C_t|$$

is a perfect square. When $\lambda_1 = \lambda_2$, then this case reduces to that of a b.i.b.d., considered in Chapter II (page 20). To obtain the same formulae as in that case, note that $rk - v\lambda = r - \lambda$, and that the matrix C_t defined in the present chapter reduces to $k(r-\lambda)$ times that in Chapter II. Hence the two determinants $|C_t|$ are in the ratio of $k^t(r-\lambda)^t$.

We have seen in the last chapter (see also [54]) that if the incidence matrix of a symmetric p.b.i.b.d. is not singular, and if the greatest common divisor of certain determinants there defined is unity, then the

---

*Remember that

$$\begin{vmatrix} a_{11} + a_{12} + a_{13} & a_{11}a_{21} + a_{12}a_{22} + a_{13}a_{23} \\ a_{11}a_{21} + a_{12}a_{22} + a_{13}a_{23} & a_{21}^2 + a_{22}^2 + a_{23}^2 \end{vmatrix} =$$

$$\begin{vmatrix} a_{11} & a_{12} \\ a_{21} & a_{22} \end{vmatrix}^2 + \begin{vmatrix} a_{11} & a_{13} \\ a_{21} & a_{23} \end{vmatrix}^2 + \begin{vmatrix} a_{12} & a_{13} \\ a_{22} & a_{23} \end{vmatrix}^2 \geqslant 0, \quad \text{and similarly}$$

for more rows and columns.

dual of the design is a p.b.i.b.d. with the same parameters. For a GD-design the determinants in question are those of the sub-matrices of

$$\begin{pmatrix} 1 & 1 \\ \lambda_1 & \lambda_2 \\ r + \lambda_1 p_{11}^1 & \lambda_2 p_{12}^2 \\ \lambda_2 p_{22}^1 & r + \lambda_1 p_{12}^2 + \lambda_2 p_{22}^2 \end{pmatrix} = \begin{pmatrix} 1 & 1 \\ \lambda_1 & \lambda_2 \\ r + \lambda_1(n_1 - 1) & \lambda_2 n_1 \\ \lambda_2 n_2 & r + \lambda_1 n_1 + \lambda_2 p_{22}^2 \end{pmatrix}.$$

The determinant of the first two rows is $\lambda_2 - \lambda_1$, and that of the first and third rows is $(\lambda_2 - \lambda_1)n_1 + \lambda_1 - r$. In a symmetric design, $r^2 - r = \lambda_1 n_1 + \lambda_2 n_2$, so that the latter determinant reduces to $(\lambda_1 - \lambda_2) - (r^2 - v\lambda_2)$. This is relatively prime to $\lambda_2 - \lambda_1$, if $r^2 - v\lambda_2$ is. The same result for GD-designs was, in a different form, proved by Connor in [37].

Shrikhande has shown in [115] that if a is an odd factor of m, and if a symmetric regular GD-design is constructed from a difference set such that multiples of m appear λ_1 times as differences, and the other non-zero values λ_2 times, then $x^2 = (k^2 - v\lambda_2)y^2 + (-1)^{(a-1)/2}az^2$ has a solution in integers which are not all zero. If c is an odd factor of $n = v/m$, and c and m are relatively prime, then a symmetric GD-design with $k > \lambda_1$ can be constructed from a difference set as described above only if $x^2 = (k - \lambda_1)y^2 + (-1)^{(c-1)/2}cz^2$ has such a solution. (Compare similar results for b.i.b.d's in Chapter II.)

In a semi-regular GD-design we have $rk - v\lambda_2 = (r - \lambda_2) + (n-1)(\lambda_1 - \lambda_2) = 0$, i.e. $r - \lambda_1 = n(\lambda_2 - \lambda_1)$, and hence $\lambda_2 > \lambda_1$.

We now have $|AA'| = 0$. If we omit from AA' the $2n$th, $3n$th ... mnth rows and columns (but not the nth), we obtain a product of rk and m matrices, but apart from the first of these they are now only of order $n - 1$. Proceeding in a similar manner to that applied before, we find that the determinant of the remaining matrix equals

$$rk(r - \lambda_1)^{n-1}(rk - v\lambda_2 - \lambda_1 + \lambda_2)^{m-1}(r - \lambda_1)^{(m-1)(n-2)}$$
$$= rk(r - \lambda_1)^{mn - 2m + 1}(\lambda_2 - \lambda_1)^{m-1}$$

which is not zero. Hence the rank of AA' is at least $v - (m-1)$, and because it cannot exceed b, we have for semi-regular GD-designs $b \geqslant v + r - m$.

If e_j varieties of the first group of a GD association scheme appear in the jth block of the design $(j = 1, \ldots, b)$, then $\sum_{j=1}^{b} e_j = nr = bk/m$

(since $v = mn$). Further, because every pair of varieties trom the first group appears together in λ_1 blocks, we have $\sum_{j=1}^{b} \binom{e_j}{2} = \lambda_1 \binom{n}{2}$.

From this, and from $r - \lambda_1 = n(\lambda_2 - \lambda_1)$ (see above), we have in a semi-regular GD-design $\sum_{j=1}^{b} e_j^2 = \lambda_1(n-1)n + nr = n^2\lambda_2$, and hence

$$\sum_{j=1}^{b} \left(e_j - \sum_j e_j/b \right)^2 = n^2\lambda_2 - bk^2/m^2 = \frac{n}{m}(v\lambda_2 - bk^2/nm)$$

$$= \frac{n}{m}(v\lambda_2 - rk) = 0.$$

But a sum of squares cannot be zero unless all its terms vanish, and therefore $e_j = k/m$ for all j. The ratio k/m must, of course, be an integer. This applies not only to the first group, but to all of them, i.e. all blocks of a semi-regular GD-design contain the same number of varieties from all groups.

(B) Triangular association scheme

If the number of varieties is of the form $\binom{n}{2}$, then a so-called "triangular association scheme" can be constructed. We denote the varieties by a_{ij} $(i,j=1,\ldots,n,\ i < j)$, and write the scheme

$$
\begin{array}{ccccc}
\cdot & a_{12} & a_{13} & \cdots & a_{1n} \\
a_{12} & \cdot & a_{23} & \cdots & a_{2n} \\
& - & - & - & \\
a_{1n} & a_{2n} & a_{3n} & \cdots & \cdot
\end{array}
$$

First associates are pairs of varieties in the same columns, and all other pairs are second associates. Every variety appears in two columns and therefore has $n_1 = 2(n-2)$ first associates. Hence $n_2 = \binom{n-2}{2}$. If two varieties are first associates, then they appear once in the same column, so that $p_{11}^1 = n-2$. From this we obtain

$$(p_{jk}^1) = \begin{pmatrix} n-2 & n-3 \\ n-3 & \binom{n-3}{2} \end{pmatrix} ; \quad (p_{jk}^2) = \begin{pmatrix} 4 & 2(n-4) \\ 2(n-4) & \binom{n-4}{2} \end{pmatrix}.$$

If $n = 3$, then we have a b.i.b.d.

We give an example of a p.b.i.b.d. with such an association scheme, where $n = 6$, $\lambda_1 = 0$, $\lambda_2 = 1$.

1	10	15		2	7	15		3	8	12
1	11	14		2	8	14		3	6	15
1	12	13		2	9	13		3	9	11

4	6	14		5	6	13
4	7	12		5	7	11
4	9	10		5	8	10

This is the design mentioned in [19] and referred to on page 75.

Conditions for the existence of such designs are given in [78] and [79]. (The latter deals also with other types of designs.) Again, the association scheme does not define the design. In [19] four designs are given for the same triangular scheme, namely that for $v = 10$.

W.S. Connor has shown that if in a scheme with two associate classes $n_1 = 2(n-2)$ holds, and if the parameters of the second kind have the values given above, then the association scheme is triangular, provided $n \geqslant 9$. Shrikhande has shown the same for $n \leqslant 6$ (see [38], [117] and [30]), and Hoffman has proved in [52] that it is also true for $n = 7$, but not for $n = 8$. He has also described, in [53], how to enumerate all counter-examples for $n = 8$. A complete list of these is contained in [31].

If in a p.b.i.b.d. with a triangular association scheme the equality $2(rk - v\lambda_1) = n(r - \lambda_1)$ holds, which is the same as $2rk - nr - n(n-2)\lambda_1 = 0$, then $2k$ is divisible by n, and every block contains precisely $2k/n$ varieties from each of the n rows of the association scheme. This is proved in [89] as follows:

Let e_{ij} be the number of varieties in block j from row i of the scheme. Then

$$\sum_{j=1}^{b} e_{ij} = r(n-1), \quad \sum_{j=1}^{b} \binom{e_{ij}}{2} = \lambda_1 \binom{n-1}{2}, \qquad \text{so that}$$

$$\sum_{j=1}^{b} e_{ij}^2 = (n-1)\left[r + \lambda_1(n-2)\right].$$

We have $\quad \bar{e}_i = \sum_{j=1}^{b} e_{ij}/b = r(n-1)/b = 2k/n, \qquad$ and

$$\sum_{j=1}^{b} (e_{ij} - \bar{e}_i)^2 = (n-1)\left[r + \lambda_1(n-2)\right] - 4bk^2/n^2 =$$

$$= \frac{n-1}{n}[nr + \lambda_1 n(n-2)] - 4vrk/n^2 = \frac{(n-1)}{n}[nr + \lambda_1 n(n-2) - 2rk]$$

$$= 0.$$

Hence $e_{ij} = \bar{e}_i = 2k/n$ for all i. This proves the statement.

For the special case $b = \binom{n-1}{2}$, $k=n$, $r=n-2$, $\lambda_1 = 1$, $\lambda_2 = 2$ this was proved, by a different method, in [110]. In that case $2k/n = 2$.

A computation similar to that for the GD-design shows that for a triangular design

$$|AA'| = rk(r-2)(\lambda_1+\lambda_2)^{\frac{1}{2}n(n-3)}[r + (n-4)\lambda_1 - (n-3)\lambda_2]^{n-1}.$$

(C) Latin-square type designs

Consider an orthogonal array $[n^2, c, n, 2]$, equivalent to $c-2$ orthogonal Latin squares of side n when $c > 2$. Label the columns from 1 to n^2 and let these labels be the varieties of a design. Two varieties are first associates if their columns contain, in any row, the same element; otherwise they are second associates.

Example [25, 4, 5, 2]

1	2	3	4	5	6	7	8	9	10	11	12	13	14	15	16	17	18	19	20	21	22	23	24	25
0	0	0	0	0	1	1	1	1	1	2	2	2	2	2	3	3	3	3	3	4	4	4	4	4
0	1	2	3	4	0	1	2	3	4	0	1	2	3	4	0	1	2	3	4	0	1	2	3	4
0	1	2	3	4	1	2	3	4	0	2	3	4	0	1	3	4	0	1	2	4	0	1	2	3
0	1	2	3	4	2	3	4	0	1	4	0	1	2	3	1	2	3	4	0	3	4	0	1	2

Second associates of 1 are $7, 8, 13, 15, 17, 19, 24, 25$, and all other varieties are its first associates.

An association scheme thus defined is said to be of "Latin-square type" and is denoted by L_c. L_2, in particular, can be described equivalently as an arrangement of n^2 varieties in a square of side n, where first associates are any two varieties either in the same row, or in the same column. This scheme has already been mentioned in Chapter IV (page 61) as a modification of the rectangular association scheme.

To determine n_1, take any column and its topmost element. There will be $n-1$ other columns with the same top entry. This is also true of the 2nd, 3rd,..., cth entry in the column. Since $\lambda_1 = 1$, there are $n_1 = (n-1)c$ columns that correspond to first associates. Hence $n_2 = (n^2 - 1) - (n-1)c = (n-1)(n+1-c)$.

To determine the parameters of the second kind, we first find p^1_{11}. Two columns of the array correspond to first associates if in any of the c rows they contain the same element. Without loss of generality we may assume that we choose two varieties for which this is so in the first row.

We must now find all those other columns which correspond to first associates of both varieties in the pair chosen. All those further columns, $n-2$ in number, which also have the same element in the first row, are first associates of both.

Take any of the $\binom{c-1}{2}$ pairs of rows from the array, ignoring the first row. Assume that the columns of the original pair read, in such a pair of rows, $\binom{a_1}{a_2}\binom{b_1}{b_2}$. A column with $\binom{a_1}{b_2}$ in the same two rows will correspond to a first associate of the variety with $\binom{a_1}{a_2}$ by virtue of a_1, and to a first associate of that of $\binom{b_1}{b_2}$ by virtue of b_2. Hence this column must be counted when enumerating p_{11}^1, as must also that with $\binom{b_1}{a_2}$.

Accounting for all $\binom{c-1}{2}$ pairs of rows apart from the first, we thus have $2\binom{c-1}{2}$ more columns, so that altogether $p_{11}^1 = 2\binom{c-1}{2} + n-2 = n+c(c-3)$. From this value we obtain

$$(p_{jk}^1) = \begin{pmatrix} n+c(c-3) & (c-1)(n-c+1) \\ (c-1)(n-c+1) & (n-c)(n-c+1) \end{pmatrix}$$

$$(p_{jk}^2) = \begin{pmatrix} c(c-1) & c(n-c) \\ c(n-c) & (n-c)^2 + (c-2) \end{pmatrix}.$$

It should be noted that reversing the roles of first and second associates changes these matrices from those for c into those for $n-c+1$.

We know that if n is the power of a prime, then c can be as large as $n+1$. Then $n_2 = 0$, i.e. all varieties are first associates.

For $c = 1$, the parameters of the second kind are

$$(p_{jk}^1) = \begin{pmatrix} n-2 & 0 \\ 0 & n(n-1) \end{pmatrix}, \qquad (p_{jk}^2) = \begin{pmatrix} 0 & n-1 \\ n-1 & n(n-2) \end{pmatrix}.$$

This is a GD-design, and so is, by the foregoing remark, that of $c = n$.

For $c = 2$, we have

$$(p_{jk}^1) = \begin{pmatrix} n-2 & n-1 \\ n-1 & (n-1)(n-2) \end{pmatrix}, \quad (p_{jk}^2) = \begin{pmatrix} 2 & 2(n-2) \\ 2(n-2) & (n-2)^2 \end{pmatrix}.$$

From this we can obtain the parameters for $c = n-1$.

Raghavarao has proved, in [89], that if $rk - v\lambda_1 = n(r - \lambda_1)$, then every block contains k/n varieties from each of the n rows or columns of the association scheme. The proof is analogous to that of the similar theorem for triangular association schemes.

For a Latin square design, similar calculations to those which we have carried out for GD-designs lead to

$$|AA'| = rk[r - (c-n)(\lambda_1 - \lambda_2) - \lambda_2]^{c(n-1)} [r - c(\lambda_1 - \lambda_2) - \lambda_2]^{(n-1)(n-c+1)}.$$

Shrikhande has shown in [118] that when $n \geqslant 2$, but not 4, and $n_1 = 2n-2$, $p_{11}^1 = n-2$, $p_{11}^2 = 2$, then the association scheme must be L_2. If $n = 4$, then a corresponding L_2 exists, but so does another association scheme with those parameters. D.M. Mesner has given corresponding results for L_c, $c > 2$ (see [118] and, for large n, a thesis mentioned in [38]).

If we construct a design by putting into the same block those varieties which have the same element in the first row (there will be n such blocks), then those which have the same element in the second row, etc., we obtain a p.b.i.b.d. with $v = n^2$, $b = cn$, $r = c$, $k = n$, $\lambda_1 = 1$, $\lambda_2 = 0$. For $c = n+1$, we have a b.i.b.d., equivalent to a Euclidean plane. For $c = n-1$, we have, after reversing the roles of first and second associates, the parameters of an L_2 scheme, because $n - (n-1) + 1 = 2$, and provided that n is not 4, we must, by Shrikhande's theorem, be able to write the varieties into a square scheme in such a way that any two in the same row or in the same column are first associates.

If we then construct $2n$ new blocks additional to the original scheme by writing into the same block those varieties which are in the same row, or in the same column, of the square scheme, and then incorporate these blocks into the orthogonal array $[n^2, n-1, n, 2]$ we obtain an $[n^2, n+1, n, 2]$. We have thus proved, following Shrikhande (see [120]), that the existence of a set of $n-3$ orthogonal Latin squares of side $n > 4$ implies the existence of a complete set of $n-1$ orthogonal Latin squares of side n, and thus that of a $PG(2,n)$.

Example

In the [25, 4, 5, 2] given above, we reverse the roles of first and second associates. The association scheme can then be written

1	25	19	13	7
8	2	21	20	14
15	9	3	22	16
17	11	10	4	23
24	18	12	6	5

(Note, incidentally, that this is a magic square: all rows and columns add up to 65.)

We can then add the following two rows to the array:

0 1 2 3 4 4 0 1 2 3 3 4 0 1 2 2 3 4 0 1 1 2 3 4 0

0 1 2 3 4 3 4 0 1 2 1 2 3 4 0 4 0 1 2 3 2 3 4 0 1

The 6 rows are equivalent to four orthogonal Latin squares of side 5.

The following generalisation of this theorem was later proved by Bruck in [25]. If $n > x^4/2 + x^3 + x^2 + 3x/2$, $x = n - t - 2$, then a set of t pairwise orthogonal Latin squares of side n can always be completed. Also, if $n > (n - t - 2)^2$, and if the set can be completed, then the completion is unique.

In [69] C.R. Nair has given necessary and sufficient conditions for a design obtained from a p.b.i.b.d. of linked-block type either by block section, or by block intersection, to be again a p.b.i.b.d. with the same association scheme, provided the latter is rectangular, *GD*, triangular, or of L_c type.

EXERCISES

(1) Derive designs from (i) a tetrahedron, (ii) an octahedron.

(2) Construct a p.b.i.b.d. by omitting, from a $EG(2,3)$, one point and all the lines through it.

(3) The following points on a $PG(2,4)$ form a set, no three of which are collinear:

$A(1,1,z)$, $\quad B(0,1,y)$, $\quad C(1,1,y)$, $\quad D(1,z,y)$, $\quad E(1,z,z)$

(see Exercise 1 of Chapter II).

Construct p.b.i.b.d's by E. Seiden's methods (page 76).

(4) From a $EG(2,3)$, construct a b.i.b.d. with $v = 9$, $b = 12$, $k = 3$, $r = 4$, and invert it.

(5) Determine the parameters of the following group-divisible design:

$$
\begin{array}{cccc}
1_1 & 1_2 & 2_1 & 2_2 \\
1_1 & 1_2 & 3_1 & 3_2 \\
2_1 & 2_2 & 3_1 & 3_2 \; .
\end{array}
$$

SOLUTIONS TO EXERCISES

Chapter II

(1) The 15 b-points are

$$d(100) \quad a(001) \quad e(101) \quad f(10y) \quad g(10z)$$
$$k(1y1) \quad h(110) \quad n(1z0) \quad j(1y0) \quad b(010)$$
$$c(01z) \quad i(111) \quad l(1yy) \quad o(1z1) \quad m(1yz)$$

and the 6 v-lines

(1) $x_1 = 0$ (2) $x_2 = 0$ (3) $x_0 + zx_1 = 0$

(4) $x_0 + x_2 = 0$ (5) $x_0 + yx_1 + zx_2 = 0$ (6) $x_0 + x_1 + yx_2 = 0$

This gives the following b.i.b.d. :

$$\begin{array}{ccccc} 12 & 13 & 14 & 15 & 16 \\ 34 & 26 & 25 & 23 & 24 \\ 56 & 45 & 36 & 46 & 35 \end{array}$$

(2)   3567   4671   5712   6123   7234   1345   2456

(3)

$$\begin{array}{rrrrrrrr} 1 & 1 & 1 & 1 & 1 & 1 & 1 & 1 \\ 1 & 1 & 1 & -1 & 1 & -1 & -1 & -1 \\ 1 & -1 & 1 & 1 & -1 & 1 & -1 & -1 \\ 1 & -1 & -1 & 1 & 1 & -1 & 1 & -1 \\ 1 & -1 & -1 & -1 & 1 & 1 & -1 & 1 \\ 1 & 1 & -1 & -1 & -1 & 1 & 1 & -1 \\ 1 & -1 & 1 & -1 & -1 & -1 & 1 & 1 \\ 1 & 1 & -1 & 1 & -1 & -1 & -1 & 1 \end{array}$$

(4) This is a b.i.b.d. with $v = 7$, $k = 3$, $\lambda = 1$, $k - \lambda = 2$. The Diophantine equation $x^2 = 2y^2 - z^2$ has the solution $x = y = z = 1$.

Chapter III

(1)

$$\begin{array}{lll} a & b & Q \\ c & P & Z \\ R & Y & A \end{array} \quad \text{leads to} \quad \begin{array}{cccccc} 0 & 1 & 2 & 0 & 1 & 2 \\ 1 & 2 & 0 & 2 & 0 & 1 \\ 2 & 0 & 1 & 1 & 2 & 0 \end{array}$$

Chapter III (continued)

(2) $$\mathbf{uX} + \mathbf{v}$$

	$X = 1$						$X = x$			
$v =$	0	1	x	$x+1$		$v =$	0	1	x	$x+1$
$u = 0$	0	1	x	$x+1$		$u = 0$	0	1	x	$x+1$
1	1	0	$x+1$	x		1	x	$x+1$	0	1
x	x	$x+1$	0	1		x	$x+1$	x	1	0
$x+1$	$x+1$	x	1	0		$x+1$	1	0	$x+1$	x

	$X = x+1$			
$v =$	0	1	x	$x+1$
$u = 0$	0	1	x	$x+1$
1	$x+1$	x	0	1
x	1	0	$x+1$	x
$x+1$	x	$x+1$	1	0

or, in other symbols

```
0123   0123   0123
1032   2301   3201
2301   3210   1032
3210   1032   2310
```

(3)

```
01234   01234   01234   01234
12340   23401   34012   40123
23401   40123   12340   34012
34012   12340   40123   23401
40123   34012   23401   12340
```

(4)

(i)
```
012   34
340   12
123   40
234   01
401   23
```

(ii) Clearly impossible, because the fifth symbol is missing in every row.

Chapter IV

(1) $v = 12$

```
        .  12  13  14
            A   B   C
       21   .  23  24
        D       E   F
       31  32   .   34
        G   H       1
       41  42  43   .
        J   K   L
```

The blocks are the rows and the columns of this scheme.

Association scheme

	First	Second	Third	Fourth
		Associates:		
A	BCHK	IL	EFGJ	D
B	ACEL	FG	HIDJ	G
C	ABFI	EH	DGKL	J
D	EFGJ	IL	BCHK	A
E	DFBL	CJ	AKGI	H
F	DECI	BG	AHJL	K
G	HIDJ	FK	ABEL	B
H	GIAK	CJ	DFBL	E
I	GHCF	AD	CFJK	L
J	KLDG	EH	ABDG	C
K	JLAH	BG	CIDE	F
L	JKBE	AD	CFJK	I

$$(p_{jk}^1) = \begin{pmatrix} 1110 \\ 1010 \\ 1111 \\ 0010 \end{pmatrix} \qquad (p_{jk}^2) = \begin{pmatrix} 2020 \\ 0001 \\ 2020 \\ 0100 \end{pmatrix}$$

$$(p_{jk}^3) = \begin{pmatrix} 1111 \\ 1010 \\ 1110 \\ 1000 \end{pmatrix} \qquad (p_{jk}^4) = \begin{pmatrix} 0040 \\ 0200 \\ 4000 \\ 0000 \end{pmatrix}$$

Chapter IV (1) *(continued)*

For instance:

$$p_{13}^4 = 4 \quad A \ldots BCHK$$
$$D \ldots BCHK$$

$$p_{33}^2 = 2 \quad A \ldots EG\,FJ$$
$$I \ldots CK\,FJ$$

(2)

$$v = b = 6, \quad r = k = 3, \quad \lambda_1 = 2, \quad \lambda_2 = 1, \quad n_1 = n_2 = 2,$$

$$(p_{jk}^1) = \begin{pmatrix} 0 & 1 \\ 1 & 1 \end{pmatrix}, \quad (p_{jk}^2) = \begin{pmatrix} 1 & 1 \\ 1 & 0 \end{pmatrix}.$$

(3) B_0 is the identity matrix

B_1	B_2	B_3
01011000	00100101	00000010
10100100	00011010	00000001
01010010	10000101	00001000
10100001	01001010	00000100
10000101	01010010	00100000
01001010	10100001	00010000
00100101	01011000	10000000
00011010	10100100	01000000

$B_1 B_2$	$B_1 B_3$	$B_2 B_3$
02022030	00100101	01011000
20200203	00011010	10100100
02023020	10000101	01010010
20200302	01001010	10100001
20300202	01010010	10000101
02032020	10100001	01001010
30200202	01011000	00100101
03022020	10100100	00011010

$$\Sigma \lambda_i B_i = AA' = \begin{matrix} 32122101 \\ 23211210 \\ 12320121 \\ 21231012 \\ 21013212 \\ 12102321 \\ 01211232 \\ 10122123 \end{matrix}$$

Chapter IV (*continued*)

(4) (i) *abc   ade   bdf   cef*

Association scheme

	First	Second			First	Second
a	*bcde*	*f*	*d*		*abef*	*c*
b	*acdf*	*e*	*e*		*acdf*	*b*
c	*abef*	*d*	*f*		*bcde*	*a*

$$v = 6, \quad b = 4, \quad k = 3, \quad r = 2, \quad n_1 = 4, \quad n_2 = 1,$$

$$\lambda_1 = 1, \quad \lambda_2 = 0, \quad (p_{jk}^1) = \begin{pmatrix} 2 & 1 \\ 1 & 0 \end{pmatrix}, \quad (p_{jk}^2) = \begin{pmatrix} 4 & 0 \\ 0 & 0 \end{pmatrix}.$$

(ii)

$$adgj \quad bdik \quad cfgk$$
$$aehk \quad cdhl \quad bfhj$$
$$afil \quad begl \quad ceij$$

$$v = 12, \quad b = 9, \quad k = 4, \quad r = 3$$
$$n_1 = 9, \quad n_2 = 2, \quad \lambda_1 = 1, \quad \lambda_2 = 0$$

Association scheme

	First	Second			First	Second
a		*bc*	*g*			*hi*
b	*defghijkl*	*ac*	*h*	*abcdefjkl*		*gi*
c		*ab*	*i*			*gh*
d		*ef*	*j*			*kl*
e	*abcghijkl*	*df*	*k*	*abcdefghi*		*jl*
f		*de*	*l*			*jk*

$$(p_{jk}^1) = \begin{pmatrix} 6 & 2 \\ 2 & 0 \end{pmatrix} \quad (p_{jk}^2) = \begin{pmatrix} 9 & 0 \\ 0 & 1 \end{pmatrix}.$$

(5)

a	678	*d*	69E	*g*	89X
b	79X	*e*	6XE	*h*	689
c	8XE	*f*	79E	*i*	78E
		j	67X		

$$v = 6, \quad b = 10, \quad r = 5, \quad k = 3, \quad \lambda = 2$$

Chapter IV (5) (*continued*)

Inverted: *adehj bdfgh abfij bcegj acghi cdefi*

$v = 10$, $b = 6$, $r = 3$, $k = 5$, $n_1 = 3$, $n_2 = 6$ $\lambda_1 = 2$, $\lambda_2 = 1$

Association scheme

	First		First
a	hij	f	bdi
b	fgj	g	bch
c	egi	h	adg
d	efh	i	acf
e	cdj	j	abe

$$(p_{jk}^1) = \begin{pmatrix} 0 & 2 \\ 2 & 4 \end{pmatrix} \quad (p_{jk}^2) = \begin{pmatrix} 1 & 2 \\ 2 & 3 \end{pmatrix} .$$

(6) First: 14 24 34   This is a $(6_3, 9_2)$
     15 25 35
     16 26 36

Association scheme (1st associates):

$\left.\begin{matrix} 1 \\ 2 \\ 3 \end{matrix}\right\}$ 4 5 6 $\left.\begin{matrix} 4 \\ 5 \\ 6 \end{matrix}\right\}$ 1 2 3

Inverted: 123 147 456 258 789 369

$n_1 = 4$, $n_2 = 4$, $\lambda_1 = 1$, $\lambda_2 = 0$

$$(p_{jk}^1) = \begin{pmatrix} 1 & 2 \\ 2 & 2 \end{pmatrix}, \quad (p_{jk}^2) = \begin{pmatrix} 2 & 2 \\ 2 & 1 \end{pmatrix}.$$

Second: 2 7 12   4 5 10
        3 8 10   1 6 12
        4 6 11   2 8 9   This is a
        1 8 11   3 6 9 (12_3).
        4 7 9   2 5 11
        3 5 12   1 7 10

Chapter IV (6) (*continued*)

Association scheme

	First	Second	Third
1	6 7 8 10 11 12	2 3 4	5 9
2	5 7 8 9 11 12	1 3 4	6 10
3	5 6 8 9 10 12	1 2 4	7 11
4	5 6 7 9 10 11	1 2 3	8 12
5	2 3 4 10 11 12	6 7 8	1 9
6	1 3 4 9 11 12	5 7 8	2 10
7	1 2 4 9 10 12	5 6 8	3 11
8	1 2 3 9 10 11	5 6 7	4 12
9	2 3 4 6 7 8	10 11 12	1 5
10	1 3 4 5 7 8	9 11 12	2 6
11	1 2 4 5 6 8	9 10 12	3 7
12	1 2 3 5 6 7	9 10 11	4 8

Chapter V

(1)   (i)   123   124   134   234   (A b.i.b.d.)

$$v = b = 4, \quad k = r = 3, \quad \lambda = 2.$$

(ii) 134   135   146   156   235   234   256   246

$$v = 6, \quad b = 8, \quad k = 3, \quad r = 4, \quad \lambda_1 = 2, \quad \lambda_2 = 0, \quad n_1 = 4, \quad n_2 = 1$$

$$(p_{jk}^1) = \begin{pmatrix} 2 & 1 \\ 1 & 0 \end{pmatrix}, \qquad (p_{jk}^2) = \begin{pmatrix} 4 & 0 \\ 0 & 0 \end{pmatrix}.$$

(2)   Omit (0,0). The remaining points are

$a(0,1)$ $b(0,2)$ $c(1,0)$ $d(1,1)$ $e(1,2)$ $f(2,0)$ $g(2,1)$ $h(2,2)$

The remaining lines (blocks) are

$x_1 = 1 \ldots cde$ $x_2 = 1 \ldots adg$ $x_1 + x_2 = 1 \ldots ach$ $x_1 + 2x_2 = 1 \ldots bcg$ $x_1 = 2 \ldots fgh$ $x_2 = 2 \ldots beh$ $x_1 + x_2 = 2 \ldots bdf$ $x_1 + 2x_2 = 2 \ldots aef$

Association scheme:

The following pairs are second associates:

$a-b, \quad c-f, \quad d-h, \quad e-g.$

(3)   The points $A, B, C, D,$ and E form a conic, and their nucleus is $F(0,1,1)$ (cf. page 4). Let the remaining 15 points be denoted

by l.c. letters as in Exercise II.1. Then we have the following 15 lines, each of which contains two of $A, B, \ldots, F$ and three other points:

$x_0 + yx_1 + x_2 = 0$	$ABeln$	$x_0 = 0$	$BFabc$
$x_0 + x_1 = 0$	$ACahi$	$x_0 + zx_2 = 0$	$CDbfl$
$x_1 + yx_2 = 0$	$ADcdk$	$x_0 + zx_1 + x_2 = 0$	$CEcej$
$x_0 + yx_2 = 0$	$AEbgm$	$x_0 + yx_1 + yx_2 = 0$	$CFgkn$
$x_0 + zx_1 + zx_2 = 0$	$AFfjo$	$x_0 + yx_1 = 0$	$DEano$
$x_1 + zx_2 = 0$	$BCdmo$	$x_0 + x_1 + x_2 = 0$	$DFehm$
$x_0 + zx_1 + yx_2 = 0$	$BDgij$	$x_1 + x_2 = 0$	$EFdil$
$x_0 + x_1 + zx_2 = 0$	$BEfhk$		

This is the first design.

Second design:

6 points with 5 points on each line:

$x_1 = 0$	$adefg$	$x_2 = 0$	$bdh\jmath n$
$x_0 + x_2 = 0$	$beiko$	$x_0 + zx_1 = 0$	$a\jmath klm$
$x_0 + x_1 + yx_2 = 0$	$cghlo$	$x_0 + yx_1 + zx_2 = 0$	$cfimn$

(4) The b.i.b.d. is

$$
\begin{array}{cccc}
123 & 145 & 167 & 189 \\
468 & 269 & 258 & 247 \\
579 & 378 & 349 & 356
\end{array}
$$

Inverting it, we obtain

$$
\begin{array}{cccc}
1\ 7\ \ 8\ \ 9 & 3\ 6\ 8\ 11 & 1\ 4\ 6\ 10 & 1\ 5\ 11\ 12 \\
2\ 5\ \ 8\ 10 & 3\ 4\ 5\ \ 9 & 2\ 4\ 7\ 11 & 2\ 6\ \ 9\ 12 \\
3\ 7\ 10\ 12 & & &
\end{array}
$$

$$
n_1 = 9, \quad n_2 = 2 \quad \lambda_1 = 1, \quad \lambda_2 = 0 \qquad (p^1_{jk}) = \begin{pmatrix} 6 & 2 \\ 2 & 0 \end{pmatrix},
$$

$$
(p^2_{jk}) = \begin{pmatrix} 9 & 0 \\ 0 & 1 \end{pmatrix}.
$$

This is a group-divisible p.b.i.b.d., with the groups as follows (after exchanging first and second associates):

$$
1\ 2\ 3 \quad 4\ 8\ 12 \quad 5\ 6\ 7 \quad 9\ 10\ 11.
$$

Chapter V (continued)

(5) $m = 3, \quad n = 2, \quad v = 6, \quad b = 3, \quad r = 2, \quad k = 4$

$n_1 = 1, \quad n_2 = 4, \quad \lambda_1 = 2, \quad \lambda_2 = 1.$

$$A = \begin{pmatrix} 1 & 1 & 0 \\ 1 & 1 & 0 \\ 1 & 0 & 1 \\ 1 & 0 & 1 \\ 0 & 1 & 1 \\ 0 & 1 & 1 \end{pmatrix} \quad \text{and hence} \quad |AA'| = 0.$$

Also, $rk(r - \lambda_1)^{m(n-1)}(rk - v\lambda_2)^{m-1} = 0.$

BIBLIOGRAPHY

[1] Berman, G., A three parameter family of partially balanced incomplete block designs with two associate classes. *Proc. Amer. Math. Soc.*, **6** (1955), 490–493

[2] Bhattacharya, K.N., A new solution in symmetrical balanced incomplete block designs $(v = b = 31,\ r = k = 10,\ \lambda = 3)$. *Sankhyā*, **7** (1946), 423–424

[3] Bose, R.C., On the application of the properties of Galois fields to the construction of hyper-Graeco-Latin squares. *Sankhyā*, **3** (1938), 323–338

[4] ——————, On the construction of balanced incomplete block designs. *Ann. Eugenics*, **9** (1939), 353–399

[5] ——————, On some new series of balanced incomplete block designs. *Bull. Calcutta Math. Soc.*, **34** (1942), 17–31

[6] ——————, A note on Fisher's inequality for balanced incomplete block designs. *Ann. Math. Statist.*, **20** (1949), 619–620

[7] ——————, A note on orthogonal arrays (abstract). *Ann. Math. Statist.*, **21** (1950), 304–305

[8] ——————, Partially balanced incomplete block designs with two associate classes involving only two replications. *Calcutta Statist. Ass. Bull.*, **3** (1951), 120–125

[9] ——————, On the application of finite projective geometry for deriving a certain series of balanced Kirkman arrangements. *Calcutta Math. Soc. Golden Jubilee Commemoration volume (1958–1959)*, 1959, 341–354

[10] ——————, Strongly regular graphs, partial geometries and partially balanced designs. *Pacific J. Math.*, **13** (1963), 389–419

[11] Bose, R.C. and Bush, K.A., Orthogonal arrays of strength 2 and 3. *Ann. Math. Statist.*, **23** (1952), 508–524

[12] Bose, R.C. and Clatworthy, W.H., Some classes of partially balanced incomplete block designs. *Ann. Math. Statist.*, **26** (1955), 212–232

[13] Bose, R.C., Clatworthy, W.H. and Shrikhande, S.S., *North Carolina Agric. Exp. Station Techn. Bull. no. 107* (1954)

[14] Bose, R.C. and Connor W.S., Combinatorial properties of group divisible incomplete block designs. *Ann. Math. Statist.*, **23** (1952), 367–383

[15] Bose, R.C. and Mesner, D.M., On linear association algebras corresponding to association schemes of partially balanced designs. *Ann. Math. Statist.*, **30** (1959), 21–38

[16] Bose, R.C. and Nair, K.R., Partially balanced incomplete block designs. *Sankhyā*, **4** (1939), 337–372

[17] —————— , —————— , On complete sets of Latin squares. *Sankhyā*, **5** (1941), 361–382

[18] —————— , —————— , Resolvable incomplete block designs with two replications. *Sankhyā*, A**24** (1962), 9–24

[19] Bose, R.C. and Shimamoto, T., Classification and analysis of partially balanced incomplete block designs with two associate classes. *J. Amer. Statist. Ass.*, **47** (1952), 151–184

[20] Bose, R.C. and Shrikhande, S.S., On the falsity of Euler's conjecture about the non-existence of two orthogonal Latin squares of order $4t + 2$. *Proc. Nat. Acad. Sci. U.S.A.*, **45** (1959), 734–737

[21] —————— , —————— , On the construction of sets of mutually orthogonal Latin squares and the falsity of a conjecture of Euler. *Trans. Amer. Math. Soc.*, **95** (1960), 191–209

[22] Bose, R.C., Shrikhande, S.S. and Bhattacharya, K.N., On the construction of group divisible incomplete block designs. *Ann. Math. Statist.*, **24** (1953), 167–195

[23] Bose, R.C., Shrikhande, S.S. and Parker, E.T., Further results on the construction of mutually orthogonal Latin squares and the falsity of Euler's conjecture. *Can. J. Math.*, **12** (1960), 189–203

[24] Brownlee, K.A. and Loraine, P.K., The relationship between finite groups and completely orthogonal squares, cubes and hypercubes. *Biometrika*, **35** (1948), 277–282

[25] Bruck, R.H., Finite nets II: Uniqueness and embedding. *Pacific J. Math.*, **13** (1963), 421–457

[26] Bruck, R.H. and Ryser, H.J., The non-existence of certain finite

102

projective planes. *Can. J. Math.*, 1 (1949), 88–93

[27] Bush, K.A., Orthogonal arrays of index unity. *Ann. Math. Statist.*, 23 (1952), 426–434

[28] ——————— , A generalisation of a theorem due to MacNeish. *Ann. Math. Statist.*, 23 (1952), 293–295

[29] Carmichael, R.D., *Introduction to the Theory of Groups of Finite Order.* Ginn & Co., Boston, 1937

[30] Chang, L.C., The uniqueness and non-uniqueness of the triangular association scheme. *Science Record*, 3 (1959), 604–613

[31] ——————— , Association schemes of partially balanced designs with parameters $v = 28$, $n_1 = 12$, $n_2 = 15$ and $p_{11}^2 = 4$. *Science Record*, 4 (1960), 12–18

[32] Chowla, S., Erdös, P. and Straus, E.G., On the maximal number of pairwise orthogonal Latin squares of a given order. *Can. J. Math.*, 2 (1950), 92–99

[33] Chowla, S. and Ryser, H.J., Combinatorial problems, *Can. J. Math.*, 2 (1950), 92–99

[34] Clatworthy, W.H., A geometrical configuration which is a partially balanced incomplete block design. *Proc. Amer. Math. Soc.*, 5 (1954), 47–55

[35] ——————— , Contributions on partially balanced block designs with two associate classes. *Nat. B. Standards Appl. Math. Series no. 47* (1956)

[36] Connor, W.S., On the structure of balanced incomplete block designs. *Ann. Math. Statist.*, 23 (1952), 57–71

[37] ——————— , Some relations among the blocks of symmetrical group divisible designs. *Ann. Math. Statist.*, 23 (1952), 602–609

[38] ——————— , The uniqueness of the triangular association scheme. *Ann. Math. Statist.*, 29 (1958), 262–266

[39] Erdös, P. and Kaplansky, I., The asymptotic number of Latin rectangles. *Amer. J. Math.*, 68 (1946), 230–236

[40] Euler, L., Recherches sur une nouvelle espèce de carrés magiques. *Verh. Genoot. Wetensch. Vlissingen*, 9 (1782), 85–239

[41] Evans, T., Embedding incomplete Latin squares. *Amer. Math. Monthly*, 67 (1960), 958–961

[42] Fisher, K.A., A system of confounding for factors with more than two alternatives, giving completely orthogonal cubes and higher powers. *Ann. Eugenics*, **12** (1945), 283–290

[43] Fisher, R.A., An examination of the different possible solutions of a problem in incomplete blocks. *Ann. Eugenics*, **10** (1940), 52–75

[44] —————— , *The Design of Experiments*. Oliver & Boyd, Edinburgh, 1942 (3rd edn)

[45] Fisher, R.A. and Yates, F., The 6×6 Latin squares. *Proc. Camb. Phil. Soc.*, **30** (1934), 492–507

[46] —————— , —————— , *Statistical Tables for Biological, Agricultural and Medical research*. Oliver & Boyd, Edinburgh, 1953 (4th edn)

[47] Hall, M, Jr., An existence theorem for Latin squares. *Bull. Amer. Math. Soc.*, **51** (1945), 387–388

[48] —————— , A survey of combinatorial analysis. *Surveys in Applied Mathematics, IV: Some aspects of analysis and probability*. Wiley, 1958

[49] Hanani, Haim, The existence and construction of balanced incomplete block designs. *Ann. Math. Statist.*, **32** (1961), 361–386

[50] Harshbarger, B., Preliminary report on the rectangular lattices. *Biometrics Bull.*, **2** (1946), 115–119

[51] —————— , Triple rectangular lattices, *Biometrics*, **5** (1949), 1–13

[52] Hoffman, A.J., On the uniqueness of the triangular association scheme. *Ann. Math. Statist.*, **31** (1960), 492–497

[53] —————— , On the exceptional case in a characterization of the arcs of a complete graph. *IBM Res. and Dev.*, **4** (1960), 487–496

[54] —————— , On the duals of symmetric partially balanced designs. *Ann. Math. Statist.*, **34** (1963), 528–531

[55] Husain, Q.M., Symmetrical balanced incomplete block designs with $\lambda = 2$, $k = 7$ or 9. *Bull. Calcutta Math. Soc.*, **37** (1945), 115–123

[56] —————— , On the totality of the solutions for the symmetrical incomplete block design. *Sankhyā*, **7** (1945/6), 204–208

[57] Husain, Q.M., Impossibility of the symmetrical incomplete block
      design with $\lambda = 2$, $k = 7$. *Sankhyā*, **7** (1946), 317–322

[58] ——————, Structure of some incomplete block designs.
      Sankhyā, **8** (1948), 381–383

[59] ——————, A note on symmetrical balanced incomplete block
      designs with $k = 9$, $\lambda = 2$. *Bull. I.S.I.*, **38** IV (1961), 11–16

[60] Kishen, K., On the construction of Latin and hyper-Graeco-Latin
      cubes and hypercubes. *J. Indian Soc. Agric. Stat.*, **2** (1950),
      20–48

[61] Levi, F.W., *Finite Geometrical Systems*. Univ. of Calcutta, 1942

[62] MacNeish, H.F., Euler squares. *Ann. Math.* (ser. 2), **23** (1922),
      221–227

[63] Majumdar, K.N., On some theorems in combinatorics relating to
      incomplete block designs. *Ann. Math. Statist.*, **24** (1953),
      377–389

[64] Mann, H.B., The construction of orthogonal Latin squares. *Ann.
      Math. Statist.*, **13** (1942), 418–423

[65] ——————, On the construction of sets of orthogonal Latin
      squares. *Ann. Math. Statist.*, **14** (1943), 401–414

[66] ——————, On orthogonal Latin squares. *Bull. Amer. Math. Soc.*,
      50 (1944), 249–257

[67] Mendelsohn, N.S., Dulmadge, A.L., Johnson D.M. and Parker, E.T.,
      Construction of m.o.l. squares. Abstract 567–4, *Notices
      Amer. Math. Soc.*, **7** (1960), 208

[68] Moore, E.H., Concerning triple systems. *Math. Ann.*, **43** (1893),
      271–285

[69] Nair, C.R., On the methods of block section and block inter-
      section applied to certain p.b.i.b. designs. *Calcutta
      Statist. Ass. Bull.*, **11** (1962), 49–54

[70] Nair, K.R., Partially balanced incomplete block designs involv-
      ing only two replications. *Calcutta Statist. Ass. Bull.*,
      3 (1950), 83–86

[71] ——————, Some 3-replicate partially balanced incomplete block
      designs. *Calcutta Statist. Ass. Bull.*, **4** (1951), 39–42

[72] ——————, Rectangular lattices and partially balanced incom-
      plete block designs. *Biometrics*, **7** (1951), 145–154

[73] Nair, K.R. and Rao, C.R., Incomplete block designs for experiments

involving several groups of varieties. *Science and Culture*, **7** (1942), 625

[74] Nandi, H.K., On the relation between certain types of tactical configurations. *Bull. Calcutta Math. Soc.*, **37** (1945), 92–94

[75] —————— , Enumeration of non-isomorphic solutions of balanced incomplete block designs. *Sankhyā*, **7** (1945/6), 305–312

[76] —————— , A further note on non-isomorphic solutions of incomplete block designs. *Sankhyā*, **7** (1945/6), 313–316

[77] Norton, H.W., The 7×7 squares. *Ann. Eugen.*, **9** (1939), 269–307

[78] Ogawa, J., A necessary condition for existence of regular and symmetrical experimental designs of triangular type with partially balanced incomplete block designs. *Ann. Math. Statist.*, **30** (1959), 1063–1071

[79] —————— , On a unified method of deriving necessary conditions for existence of symmetrical partially balanced block designs of certain types. *I.S.I. Bull.*, **38** IV (1961), 43–57

[80] Ostrowski, R.T. and van Duren K.D., On a theorem of Mann on Latin squares. *Math. Comp.*, **15** (1961), 293–295

[81] Parker, E.T., Construction of some sets of mutually orthogonal Latin squares. *Proc. Amer. Math. Soc.*, **10** (1959), 946–949

[82] —————— , A computer search for Latin squares orthogonal to Latin squares of order 10. Abstract 564–71, *Notices Amer. Math. Soc.*, **6** (1959), 798

[83] —————— , Orthogonal Latin squares. *Proc. Nat. Acad. Sci. U.S.A.*, **45** (1959), 859–862

[84] —————— , Computer study of orthogonal Latin squares of order 10. *Computers and Automation*, **11** (1962), 33–35

[85] —————— , Non-extendibility conditions on mutually orthogonal Latin squares. *Proc. Amer. Math. Soc.*, **13** (1962), 219–221

[86] Plackett, R.L. and Burman, J.P., The design of optimum multifactorial experiments. *Biometrika*, **33** (1943–6), 305–325

[87] Raghavarao, D., A note on the construction of *GD* designs from hyper-Graeco-Latin cubes of the first order. *Bull. Calcutta Statist.Ass.*, **9** (1959), 67–70

[88] —————— , A generalization of group divisible designs. *Ann. Math. Statist.*, **31** (1960), 756–765

[89] —————— , On the block structure of certain p.b.i.b. designs

106

with two associate classes having triangular and L_2 associate schemes. *Ann. Math. Statist.*, **31** (1960), 787–791

[90] Rao, C.R., Hypercubes of strength "*d*" leading to confounded designs in factorial experiments. *Bull. Calcutta Math. Soc.*, **38** (1946), 67–78

[91] ———— , A general class of quasi-factorial and related designs. *Sankhyā*, **17** (1956–7), 165–174

[92] ———— , Combinatorial arrangements analogous to orthogonal arrays. *Sankhyā*, **23**A (1961), 283–286

[93] Rao, P.V., On the construction of some partially balanced incomplete block designs with more than three associate classes. *Bull. Calcutta Statist. Ass.* 9 (1959), 87–92

[94] ———— , The dual of a balanced incomplete block design. *Ann. Math. Statist.*, **31** (1960), 779–785

[95] Ray-Chaudhuri, D.K., Application of the geometry of quadrics for constructing p.b.i.b. designs. *Ann. Math. Statist.*, **33** (1962), 1175–1186

[96] Reiss, M., Über eine Steinersche kombinatorische Aufgabe, welche im 45. Band dieses Journals, Seite 181, gestellt worden ist. *Crelle's J. reine und angew. Math.*, **56** (1859), 326–344

[97] Rojas, B. and White, R.F., The modified Latin square. *J. Royal Stat. Soc.* (B), **19** (1957), 305–317

[98] Rouse-Ball, W.W., *Mathematical recreations and essays* (revised by H.S.M. Coxeter). MacMillan, London, 1942

[99] Roy, J. and Laha, R.G., Two associate partially balanced designs involving three replications. *Sankhyā*, **17** (1956–7), 175–184

[100] Roy, P.M., Hierarchical group divisible incomplete block designs with *m* associate classes. *Science and Culture*, **19** (1953–4), 210–211

[101] ———— , On the method of inversion in the construction of partially balanced incomplete block designs from the corresponding balanced incomplete block designs. *Sankhyā*, **14** (1954), 39–52

[102] ———— , On the properties and construction of *HGD* designs with *m* associate classes. *Calcutta Statist. Ass. Bull.*, **11** (1962), 10–38

[103] Ryser, H.J., A combinatorial theorem with application to Latin rectangles. *Proc. Amer. Math. Soc.*, **2** (1951), 550–552

[104] Sade, A., An omission in Norton's 7×7 squares. *Ann. Math. Statist.*, **22** (1951), 306–307

[105] Schützenberger, M.P., A non-existence theorem for an infinite family of symmetric block designs. *Ann. Eugen.*, **14** (1949), 286–287

[106] Seiden, E., A remark on the geometrical method of construction of an orthogonal array. *Ann. Math. Statist.*, **25** (1954), 177–178

[107] ————, On the maximum number of constraints of an orthogonal array. *Ann. Math. Statist.*, **26** (1955), 132–135

[108] ————, Further remark on the maximum number of constraints of an orthogonal array. *Ann. Math. Statist.*, **26** (1955), 759–763

[109] ————, On a geometrical method of construction of partially balanced designs with two associate classes. *Ann. Math. Statist.*, **32** (1961), 1177–1180

[110] ————, On necessary conditions for the existence of some symmetrical and unsymmetrical triangular p.b.i.b. designs and b.i.b. designs *Ann. Math. Statist.*, **34** (1963), 348–351

[111] Shah, B.V., On a generalization of the Kronecker product design. *Ann. Math. Statist.*, **30** (1959), 48–54

[112] Shrikhande, S.S., The impossibility of certain symmetrical balanced incomplete block designs. *Ann. Math. Statist.*, **21** (1950), 106–111

[113] ————, On the dual of some incomplete block designs. *Biometrics*, **8** (1952), 66–72

[114] ————, The non-existence of certain affine resolvable balanced incomplete block designs. *Can. J. Math.*, **5** (1953), 413–420

[115] ————, Cyclic solutions of symmetrical *GD*-designs. *Calcutta Statist. Ass. Bull.*, **5** (1953), 36–39

[116] ————, Affine resolvable balanced incomplete block designs and non-singular *GD*-designs. *Calcutta Statist. Ass. Bull.*, **5** (1954), 139–141

[117] ————, On a characterization of the triangular association scheme. *Ann. Math. Statist.*, **30** (1959), 39–47

[118] Shrikhande, S.S., The uniqueness of the L_2 association scheme. *Ann. Math. Statist.*, **30** (1959), 781–798

[119] —————————, Relations between certain incomplete block designs. *Essays in honor of Harold Hotelling.* Stanford U.P., 196C

[120] —————————, A note on mutually orthogonal Latin squares. *Sankhyā*, A, **23** (1961), 115–116

[121] Shrikhande, S.S. and Singh, N.K. On a method of constructing symmetrical balanced incomplete block designs. *Sankhyā* A, **24** (1962), 25–32

[122] Smith, H.J.S., On systems of linear indeterminate equations and congruences. *Phil. Trans. Royal Soc. London*, **151** (1961–2), 293–326

[123] Sprott, D.A., Some series of partially balanced incomplete block designs. *Can. J. Math.*, **7** (1955), 369–381

[124] —————————, A series of symmetrical group divisible incomplete block designs. *Ann. Math. Statist.*, **30** (1959), 249–250

[125] Stevens, W.L., The completely orthogonalized Latin square. *Ann. Eugen.*, **9** (1939), 82–93

[126] Tarry, G., Le problème des 36 officiers. *C.R. Acad. Franc. pour l'Avancement de Science Naturel*, **1** (1900), 122–123; **2** (1901), 170–203

[127] Thompson, W.A. Jr., A note on p.b.i.b. design matrices. *Ann. Math. Statist.*, **29** (1958), 919–922

[128] Todd, J.A., A combinatorial problem. *J. Math. Phys.*, **12** (1932–3), 321–333

[129] Vajda, S., *Patterns and Configurations in Finite Spaces*. Griffin, London, 1967

[130] Yates, F., Complex experiments. *J. Royal Stat. Soc. Suppl.*, **2** (1935), 181–247

[131] —————————, Incomplete randomized blocks. *Ann. Eugen.*, **7** (1936), 121–140

INDEX